basics of
fractional
horsepower
motors
and repair

basics of
fractional
horsepower
motors
and repair

by GERALD SCHWEITZER
Training and Service Publications Supervisor
Fedders Corporation

JOHN F. RIDER PUBLISHER, INC., NEW YORK

PREFACE

Fractional horsepower motors have long played a vital role in the operation of home appliances and equipment which everyone depends upon and takes for granted. These motors are of many types and shapes, each designed and chosen for its specific application. The purpose of this book is to provide a working explanation of the principles of operation of the various fractional horsepower motors and to present basic procedures for servicing and maintaining them.

Because of the widespread use and great number of fractional horsepower motors, proper servicing has become a problem to motor manufacturers and to manufacturers of equipment which includes motors. Many motors are returned from the field improperly tagged as to their troubles, and many more are returned to the manufacturer when they could have been repaired on the spot, had the service technician been more knowledgeable.

By answering the question "What makes this motor run?" and by examining the construction of the motor, the author believes that better maintenance and service may be rendered, eliminating unnecessary down-time and minimizing inconvenience to the ultimate user. To aid in locating faults, after each type of motor has been discussed, there is an accompanying trouble-shooting chart for that motor.

In addition to principles of operation, construction, troubleshooting, and maintenance, a section on basic motor testing describes tests that may be performed by the reader with basic test equipment. There is also a section covering enclosure and mounting characteristics. And, although the actual winding of a motor is beyond the scope of this book, common types of winding and their construction are explained.

While this book is intended for all who are interested in or concerned with the repair of fractional horsepower motors, it is also designed to help the student. For this reason, the book is profusely illustrated. This proven "picture-book" method of presentation visualizes for the reader the salient points brought out in the text and brings theory and practice into a closer relationship.

The author gratefully acknowledges the cooperation and assistance of the following people for their constructive suggestions: William P. Ruocco and Thomas J. Dolan of The General Electric Company, George I. Durfee of Westinghouse Electric Company, Murray P. Rosenthal, and Chris Behrens.

The author also wishes to express his sincere appreciation to the numerous friends and companies which contributed to the successful completion of this book.

GERALD SCHWEITZER

New York, N. Y.
February, 1960

CONTENTS

BASICS OF FRACTIONAL HORSEPOWER MOTORS

Basic Principles of Magnetism

To understand how an electric motor operates, let us first consider the basic principles of electricity and magnetism, which are related to the operation of electric motors.

The north and south poles of a permanent magnet are connected by an invisible field of magnetic force. This field can be demonstrated by placing a permanent magnet under a sheet of glass or paper and by sprinkling the

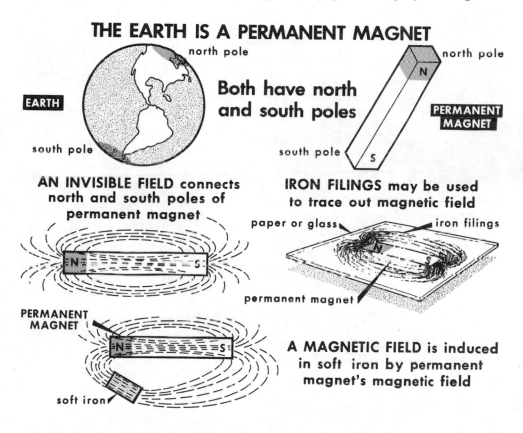

THE EARTH IS A PERMANENT MAGNET

north pole

north pole

EARTH

Both have north and south poles

PERMANENT MAGNET

south pole

south pole

AN INVISIBLE FIELD connects north and south poles of permanent magnet

IRON FILINGS may be used to trace out magnetic field

paper or glass

iron filings

permanent magnet

PERMANENT MAGNET

A MAGNETIC FIELD is induced in soft iron by permanent magnet's magnetic field

soft iron

surface of this sheet with iron filings. When the glass or paper is tapped lightly, the iron filings are attracted by the lines of force of the magnetic field and form a pattern following these lines.

If a piece of soft iron is placed within the field of a magnet, it assumes the characteristics of that magnet. Note that the piece of soft iron, when placed in the magnetic field of a permanent magnet, does not have to touch it in order to take on its characteristics. This action is called *induction*. In other words, the magnetic field of the permanent magnet *induces* a magnetic field into the piece of soft iron, causing it too to become a magnet.

Basic Principles of Magnetism — Electromagnets

If we bring two permanent magnets near each other, they either attract or repel — unlike poles attract and like poles repel. If we suspend a magnet in such a manner that it is free to turn and if we bring another magnet close to the pivoted one, the latter starts to rotate. This illustrates one of the most important principles used in the operation of electric motors, because

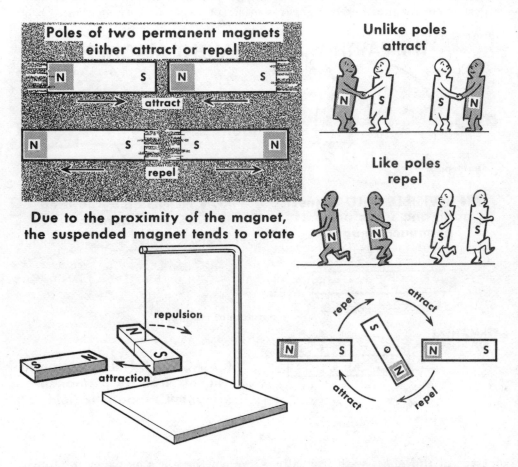

it is in a similar manner that the attracting and the repelling action of a magnetic field is converted to rotary motion in an electric motor.

A wire carrying electric current is surrounded by a magnetic field. The strength of a magnetic field around a single conductor is relatively weak. However, if we wind a coil of insulated wire around a soft-iron core, we can produce an electromagnet that is much stronger than the permanent magnet. The strength of an electromagnet is determined by the number of turns in its coil and the amount of current flowing through the coil.

Electromagnets (contd.)

An electromagnet has several other advantages over a permanent magnet. Its core is magnetized only when current flows through its coil. When no current flows, there is no magnetic field. The polarity of an electromagnet can be reversed by reversing the direction of current flow through its coil. Electromagnets perform an essential function in the electric motor — that of supplying the magnetic field necessary for its operation.

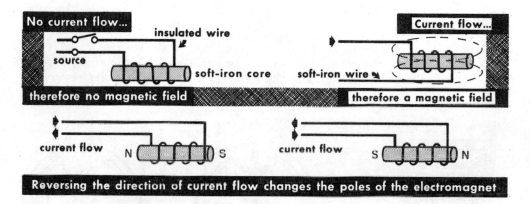

A typical *induction* motor consists of a stationary section called a *stator* and a rotating section called a *rotor*. Our stator is made up of thin sections of soft iron or steel, called laminations, which we substitute for the thin wire previously shown. For simplicity's sake, we select a stator with two poles. If we wind the soft iron laminations with insulated wire and connect them together, we have two electromagnets. If we apply an ordinary 60-cycle alternating current to our electromagnets, our stator becomes an electromagnet whose magnetic poles reverse their polarity 120 times per second, that is, every time the current direction alternates.

A TWO-POLE STATOR

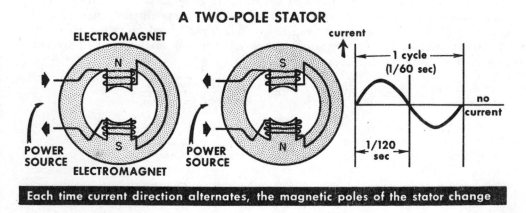

3

Electric Motor Principles

In the center of our magnetic field we mount a permanent magnet and give it a spin as we apply 60 cycles a-c to the stator coils. The permanent magnet or rotor continues to spin due to the attraction and repulsion of the alternating polarities of the stator's poles. Once started, our rotor continues to spin because, at that instant, the top pole is a north pole and the bottom pole is a south pole. The south pole of the rotor is attracted by the north pole and repelled by the south pole of the stator. Simultaneously the north pole of the rotor is attracted by the south pole and repelled by the north pole of the stator.

We thus have a push-pull action on the rotor due to the magnetic poles of the stator. Before the rotor can come to rest in line with the stator poles, the current changes direction, and the polarity of the stator poles reverses.

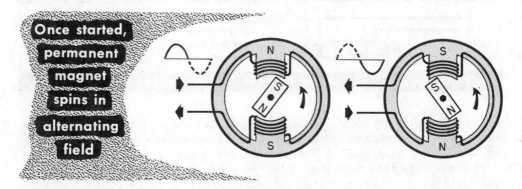

Once started, permanent magnet spins in alternating field

Momentum carries the rotor past center, and the south pole of the stator attracts the north pole of the rotor and repels its south pole. Simultaneously the north pole of the stator repels the north pole of the rotor and attracts its south pole. Thus the push-pull action is continuous. The rotor continues to spin, and the stator poles continue to reverse their polarity. The rotor theoretically adjusts itself to a speed of 60 revolutions per second (rps) or 3600 revolutions per minute (rpm). This speed is called the *synchronous* speed. The synchronous speed of a motor is determined by the use of the following formula:

$$\text{Synchronous speed} = 120 \times \frac{\text{Frequency}}{\text{Number of poles}}$$

For example:

$$\text{(for a four-pole motor)} \quad = \frac{120 \times 60 \text{ cps}}{4} = 1800 \text{ rpm}$$

$$\text{(for a six-pole motor)} \quad = \frac{120 \times 60 \text{ cps}}{6} = 1200 \text{ rpm}$$

Electric Motor Principles (contd.)

The speed of a constant-speed motor is determined by the frequency of the power supply and the number of poles. Shown below are the pole arrangements of the 4- and 6-pole motors referred to on the preceding page. The principles of their rotating field are discussed in later pages.

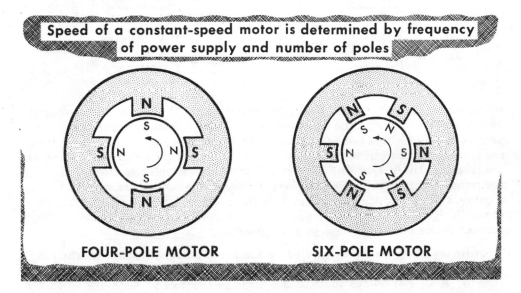

Speed of a constant-speed motor is determined by frequency of power supply and number of poles

FOUR-POLE MOTOR   SIX-POLE MOTOR

Let us review another basic principle of magnetism. When a conductor forming a closed circuit is moved through the magnetic field of a permanent magnet or an electromagnet so that the wire cuts across the lines of magnetic force, a voltage is induced in the wire and electric current flows through it. It is also true that if the lines of force of a magnet are caused to move through a stationary wire, electric current flows through the wire.

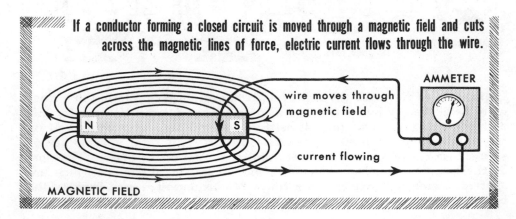

If a conductor forming a closed circuit is moved through a magnetic field and cuts across the magnetic lines of force, electric current flows through the wire.

AMMETER

wire moves through magnetic field

current flowing

MAGNETIC FIELD

Current Induced in a Coil — Rotors

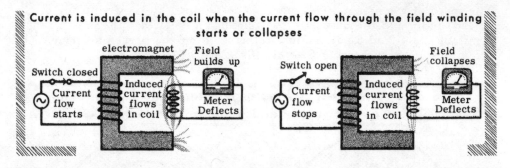

Picture a coil of wire placed between the poles of a U-shaped electromagnet. When current flows in the field winding of the electromagnet, a magnetic field is produced between the poles of the electromagnet. Its lines of force cut through the coil of wire. If the coil is connected into a closed circuit, an *induced* current is said to flow through the coil. When the current flow in the winding stops, the magnetic field disappears or collapses, again causing a current to be induced in the coil as the magnetic field again cuts across the coil. No current flows through the coil except at the very instant that current in the winding starts or stops flowing.

If ordinary 60-cycle a-c is applied to the electromagnet winding, the magnetic field builds up and collapses for every alternation, or 120 times a second. A similar 60-cps induced a-c flows in the coil of wire placed in the magnetic field. This process is called induction, as there is no direct contact between the two coils of wire.

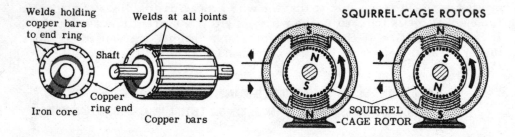

Previously we discussed the case of a permanent magnet rotating in a magnetic field. Let us now substitute a more practical type of rotor for our permanent magnet. This rotor is called a squirrel-cage rotor because of its resemblance to cages used to exercise squirrels. This rotor is composed of slotted sections of special thin steel similar to that of the stator. The slots of the rotor contain bars of bare copper, brass, or aluminum which are shorted at each end with shorting rings. No insulation is used between the bars and the rotor core since the voltage induced in these bars is very low.

Current Induced in a Coil — Rotors (contd.)

If we give this rotor a spin within our stator and apply a 60-cycle a-c to our stator coils, the expanding and collapsing stator fields induce current in the bars of the rotor as they cut the lines of magnetic force created by the stator field. These currents create a magnetic field within the rotor exactly opposite to that of the stator.

Since the rotor has been given a spin, the rotor's momentum is aided by the magnetic fields created, and our motor continues to run. Our rotor continues to increase in speed until it turns nearly 180° for each alternation of the stator field. The actual running speed of the motor is slightly less than its synchronous speed. If the rotor turns at the same speed as the magnetic

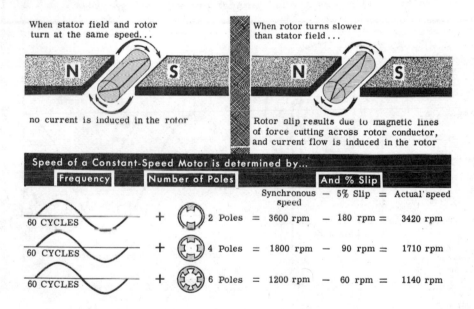

When stator field and rotor turn at the same speed...

no current is induced in the rotor

When rotor turns slower than stator field...

Rotor slip results due to magnetic lines of force cutting across rotor conductor, and current flow is induced in the rotor

Speed of a Constant-Speed Motor is determined by...

Frequency		Number of Poles			And % Slip	
				Synchronous speed	— 5% Slip =	Actual speed
60 CYCLES	+	2 Poles	=	3600 rpm	— 180 rpm =	3420 rpm
60 CYCLES	+	4 Poles	=	1800 rpm	— 90 rpm =	1710 rpm
60 CYCLES	+	6 Poles	=	1200 rpm	— 60 rpm =	1140 rpm

field of the stator, no current is induced in the rotor. Current is induced in the rotor only when its copper bars cut the lines of force created by the stator field. This condition exists only when the rotor is turning more slowly than the speed of alternations. This difference in speed is called *slip*. The greater the slip, the more lines of force are cut, and the stronger the induced current in the rotor becomes. This causes the rotor to pick up speed. As it turns faster, however, fewer lines of magnetic force are cut, the current and the number of magnetic lines of force in the rotor become weaker, and the rotor slows down slightly. The actual running speed thus becomes a balance between these two tendencies and usually runs between 4 and 5% below the synchronous speed, or about 3420 rpm. The speed of a constant-speed motor is determined by the frequency of the power supplied, the number of poles, and the percentage of slip.

Two-Phase Motors

Our motor needed a spin to start. This is because we applied a single-phase current to run our motor. Single-phase currents do not produce a natural starting torque (twist). The attraction or repulsion — the push-pull action of our stator poles — kept the motor running but was unable to start the motor. This condition can be compared to the swinging of a weight attached to a string in a circle. Once the weight is started swinging, a simple back-and-forth motion keeps it in motion; but no amount of back-and-forth

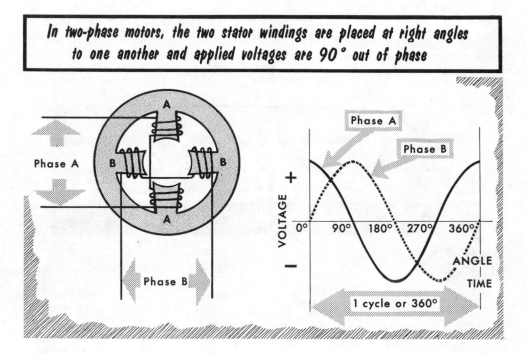

In two-phase motors, the two stator windings are placed at right angles to one another and applied voltages are 90° out of phase

motion starts it swinging in a circle. The starting requires a circular moving force. The various means used to make a single-phase motor self-starting are discussed in later pages.

Up to now we have discussed only single-phase motors. Induction motors are designed for single-, two-, and three-phase operation. In the case of the two-phase motors, the stator is arranged so that its two windings are placed at right angles to each other. When the voltages applied to phases A and B are 90° out of phase, the currents flowing in these phases are displaced by 90°. Since the magnetic fields generated by the currents are in phase with them, the fields are also displaced by 90°. Since these displaced magnetic fields have their axes at right angles to each other, they add at every instant of their cycle and produce a field which rotates one revolution for each a-c cycle.

Two-Phase Motors (contd.)

The revolving magnetic field of the stator of a two-phase motor can best be demonstrated with a chart as shown. The figure shows waveforms A and B 90° out of phase. The stator's magnetic field is shown in relation to the phases. At 0° or position 1, the current flow in phase A is at its maximum, and the current flow in phase B is zero. Therefore, the magnetic field of phase A coils in the stator is at its maximum, and the magnetic field of the phase B coils is zero. At position 2, or 45° later, the currents in both phases are equal. Thus, the magnetic fields are midway between the phase A and phase B windings. At the 90° point, or position 3, the current in phase A is zero, while the current of phase B is at maximum. Therefore, the magnetic field of the phase B coils is at its maximum while the strength of phase A coils is zero. The magnetic field has now rotated 90°.

At position 4, the strength of the magnetic fields are again equal, but the direction of the current in phase A has been reversed so that the magnetic fields are again between phases A and B but in the direction shown.

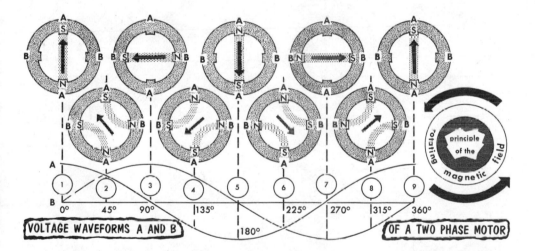

VOLTAGE WAVEFORMS A AND B

OF A TWO PHASE MOTOR

principle of the rotating magnetic field

At position 5, or 180°, the current in phase A is at its maximum but opposite in polarity from position 1, and the current in phase B is zero. Therefore, the magnetic field of the phase A coils are at their maximum, and the direction of our rotor is as shown. In positions 6 – 9, our action continues through another half-cycle to complete a revolution

Thus, we see that whenever we place two windings at right angles to each other in a stator and apply voltages that are 90° out of phase with each other, we obtain a rotating magnetic field. This rotating magnetic field pulls our rotor around with it, causing our motor to rotate. The speed with which our motor rotates is governed by the frequency of the applied voltages and the number of poles used.

The Three-Phase Motor

The operating principles of the two-phase motor apply to the three-phase motor. For the three-phase motor, however, the generated magnetic fields are 120° out of phase with each other. The windings are arranged about the stator as shown. Actually the three phases are brought to the stator by means of three wires from the three-phase power supply, as shown. The revolving magnetic field of a three-phase motor is shown in the following illustrations. Also included is a diagram of the three voltages covering one complete cycle. The numbers on the axis line refer to the numbers on the diagrams below. Each diagram shows the position of the armature at the instant indicated by the corresponding number on the curve. The action of the magnetic field is smooth and regular; the rise and fall of current in the conductors is also smooth and regular.

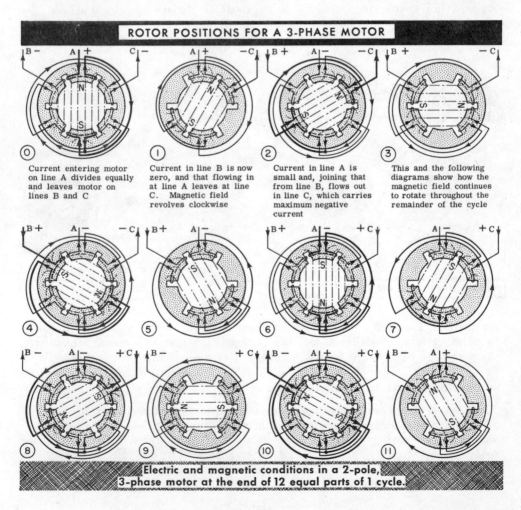

ROTOR POSITIONS FOR A 3-PHASE MOTOR

⓪ Current entering motor on line A divides equally and leaves motor on lines B and C

① Current in line B is now zero, and that flowing in at line A leaves at line C. Magnetic field revolves clockwise

② Current in line A is small and, joining that from line B, flows out in line C, which carries maximum negative current

③ This and the following diagrams show how the magnetic field continues to rotate throughout the remainder of the cycle

④ ⑤ ⑥ ⑦

⑧ ⑨ ⑩ ⑪

Electric and magnetic conditions in a 2-pole, 3-phase motor at the end of 12 equal parts of 1 cycle.

The Three-Phase Motor (contd.)

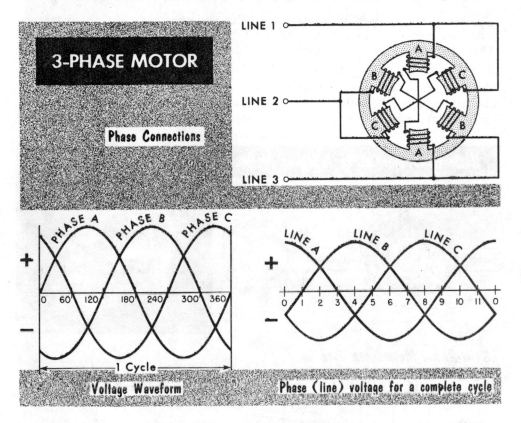

QUESTIONS AND PROBLEMS

1. What is the principle of induction?

2. How is the principle of magnetic attraction and repulsion employed in the operation of electric motors?

3. Why are soft-iron cores used for electromagnets rather than steel cores?

4. For what reason are the cores of electromagnets laminated?

5. Explain the reason a rotor continues to spin once it is started.

6. Explain the principle of operation of the squirrel-cage rotor. Why does it never reach synchronous speed? What is meant by "slip"?

7. What is the basic difference between single-phase and two-phase electric motors? Between a two- and three-phase electric motor?

8. Explain the "rotating field" of two- and three-phase electric motors.

Reviewing Some Principles of Motor Operation

The most widely used types of single-phase motors are split-phase motors. They are popular wherever motors of 1/20 to 1/3 horsepower are required. They are found in oil burners, fans, blowers, pumps, washing machines, ironers, power tools, and many other applications.

What makes a split-phase motor run? Let us review our basic theory of electric motors. A stator is needed for a motor to rotate. This stator consists of pressed steel discs (laminations), so shaped that coils can be wound

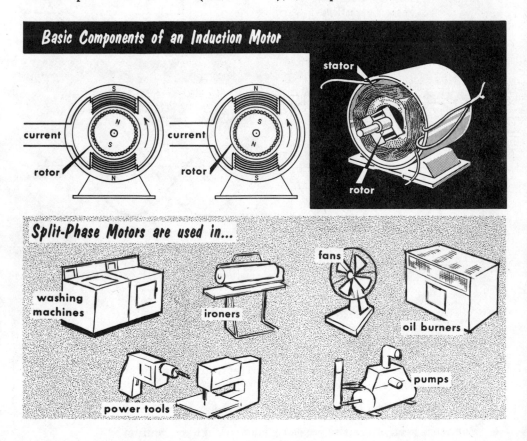

Basic Components of an Induction Motor

stator

current

rotor

current

rotor

rotor

Split-Phase Motors are used in...

washing machines

ironers

fans

oil burners

power tools

pumps

around the stator's magnetic poles. The number of poles helps to determine the motor's rotation speed. The rotor is placed within the stator whose windings are arranged to create a revolving magnetic field. If we give the rotor a spin, the motor runs. Our motor is not self-starting. Something has to be added to cause our motor to start rotating from a standstill position, that is, a *starting torque*. Torque means twist. Two-phase motors have two windings 90° apart, and it is as a result of the angular displacement of the windings that a two-phase motor develops a starting torque.

Split-Phase Motors

In split-phase motors, the starting torque is obtained by adding to our stator another winding, called a *starting winding*. It has higher resistance than the running winding. Its purpose is to obtain a phase displacement between the windings, hence the term *split-phase*. A condition similar to that of a two-phase motor is created. However, in a split-phase motor, the

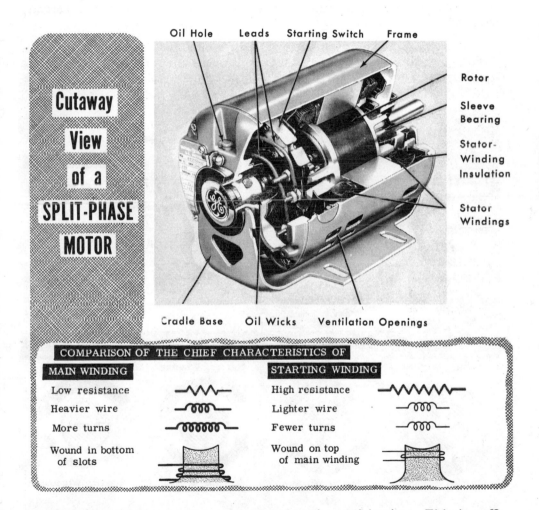

Oil Hole　Leads　Starting Switch　Frame

Rotor

Sleeve Bearing

Stator-Winding Insulation

Stator Windings

Cutaway View of a SPLIT-PHASE MOTOR

Cradle Base　Oil Wicks　Ventilation Openings

COMPARISON OF THE CHIEF CHARACTERISTICS OF

MAIN WINDING		STARTING WINDING	
Low resistance		High resistance	
Heavier wire		Lighter wire	
More turns		Fewer turns	
Wound in bottom of slots		Wound on top of main winding	

angular phase displacement rarely exceeds 20° to 30° in time. This is sufficient to provide us with enough starting torque to start the spinning of the rotor.

In actual practice, the starting winding is wound in the same slots as the running winding. The starting winding, being of higher resistance, occupies a much smaller area than the running winding.

SPLIT-PHASE MOTORS

Starting Switch

Now that our stator has two windings and we can start our motor from a dead stop, we have another problem: when our motor starts turning and comes up to approximately 75 or 80% of its synchronous speed, the main or running winding develops as much torque as the combined windings. Between 80 and 90% of the synchronous speed, our starting winding actually becomes a hindrance, and the torque developed by the motor is actually reduced.

To solve this problem, a centrifugal switch is added to the circuit of the starting winding. This mechanical switch opens the circuit of our starting winding when the rotor speed reaches approximately 80% of the synchron-

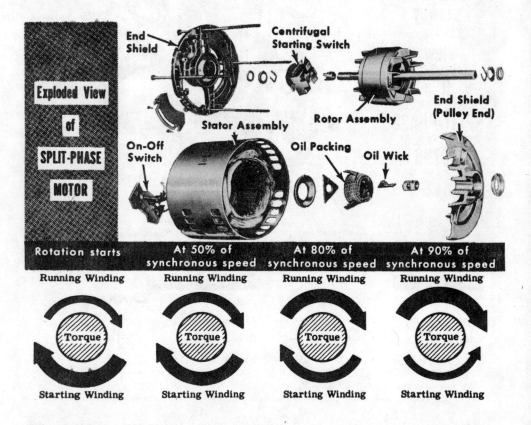

ous speed. In this manner, we obtain all the advantages of our starting winding, and when it becomes a deterrent, it is removed from the circuit. The insertion of the centrifugal switch, to cut out our starting winding from the circuit when the rotor approaches its operating speed, also prevents the motor from drawing excessive current from the line, which may burn out the high-resistance starting winding.

Split-Phase Motor Protectors

Many split-phase motor circuits include another switch to protect the motor against burnout caused by failure of the winding insulation. Damage to winding insulation is caused by overheating which may result from the lack of ventilation, high temperature, and high winding currents.

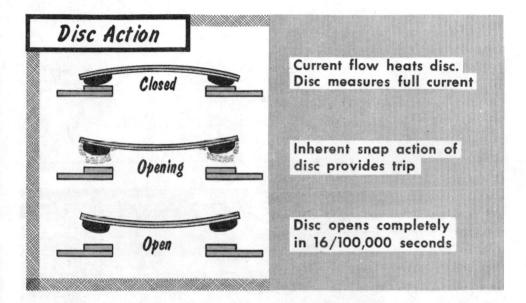

A common motor protector is the Klixon type which utilizes a temperature-responsive, snap-acting, bimetallic disc to make and break contacts. The disc is further controlled by a current-sensitive resistance heater connected in series with it. In operation, the disc is in a closed position. When a temperature rise causes the disc to snap open, the contacts are broken, and the circuit is opened. The overload protector operates as follows: The snap-action bimetallic disc is made up of two strips of dissimilar metals combined to form a single strip. The bottom strip has a greater coefficient of expansion. This means that, in case of a temperature rise, the bottom strip expands more than the top one and causes the ends to curl. This action occurs at a point where the top strip is pushing down while the bottom strip is pushing up around the corners. Then the snap-action occurs and the circuit opens. The disc is actuated by temperature rise due to the warm air surrounding the motor or to the lack of ventilation. The disc is also actuated by excessive motor current, which causes the heater element to heat up, and the bimetallic disc, in turn, expands. When the motor is shut off, the temperature returns to normal. The bimetallic disc snaps closed and restarts the motor. This type of protector constantly recycles the motor, as long as the power source remains connected, until the trouble is remedied.

Split-Phase Motor Connections

A split-phase motor consists of a running winding, a squirrel-cage rotor, a starting winding, a starting centrifugal switch, and an overheat protector. A schematic diagram of a split-phase motor with the schematic symbols used is shown below.

The methods used to terminate the leads differ with each manufacturer. In many instances, the lead arrangement is varied by a given manufacturer. Let us explain some of the more popular arrangements used.

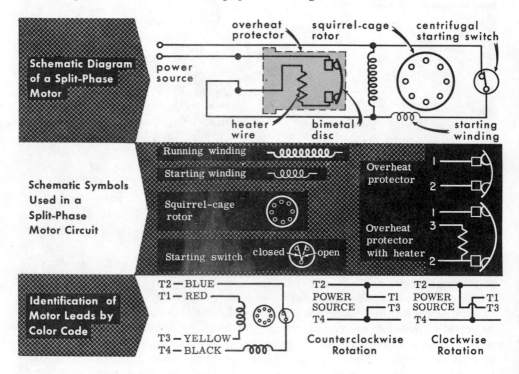

Two basic methods of tagging leads in split-phase motors are in common use today. The National Electrical Manufacturers Association has adopted a standard system of marking terminals. In this system, the leads of the running winding are tagged T1 and T3. The starting-winding leads are tagged T2 and T4.

In older motors, a system of tagging the running winding M1 and M2 and the starting winding S1 and S2 is used. In other systems, the leads are color-coded instead of being tagged. A typical color coding scheme in lieu of tagging may be as follows: T1 — Red, T2 — Blue, T3 — Yellow, and T4 — Black. This color coding is typical, but may vary greatly from the coding used in any specific motor.

Identification of Motor Leads and Windings — Direction of Rotation

Standard direction of rotation is counterclockwise as viewed from the front end

Shaft Extension   Front End   Counterclockwise

Standard Direction of Rotation

Front End

Line terminals connected for standard or counterclockwise rotation

Power Source

T2
T1
T3
T4

Counterclockwise Rotation

Line terminals connected for clockwise rotation

Power Source

T2
T1
T3
T4

Clockwise Rotation

When lead tags have been mutilated or removed, or when the colors cannot be distinguished, the leads may be identified through one or more of the following procedures.

Measure the resistance of both windings with an ohmmeter. The resistance of the running winding is almost always lower than that of the starting winding. In some special cases, the reverse may be true. If this latter condition exists, the end shield of the motor must be removed and the windings inspected. (This condition could exist in an application where an external resistance is connected to the motor.) The starting winding is, in most cases, wound on top of the running winding and is wound with smaller wire and fewer turns than the running winding. The lead from the starting winding is usually connected to the starting switch. (There can be exceptions to this rule.) If a power factor meter is available, the windings can be positively identified. (Voltmeter, ammeter, and wattmeter readings can also be used in determining power factor.) The locked (instant of starting) power factor of the starting winding is always higher than the locked power factor of the running winding when measured separately at rated voltage and frequency.

When all four leads are brought out to the line terminals, the motor may be connected so that it rotates in either a clockwise or a counterclockwise direction. The direction of motor rotation is always considered from the end opposite its shaft extension, or the front end. The standard direction of rotation is counterclockwise as viewed from the front end.

Two-Speed, Two-Winding Motors — Dual-Voltage Motors

Split-phase motors can be constructed so that two speeds are available. This is done by providing two sets of windings, one for each speed.

Split-phase motors are also convertible for use with either a 115-volt power source or a 230-volt power source merely by changing the circuit connections. Note that when the motor is connected for 115-volt operation, the two main windings are connected in parallel, thereby providing 115 volts to each winding. When the motor is connected for 230-volt operation, the two main windings are connected in series. This allows for a voltage drop across each winding to give the required voltage.

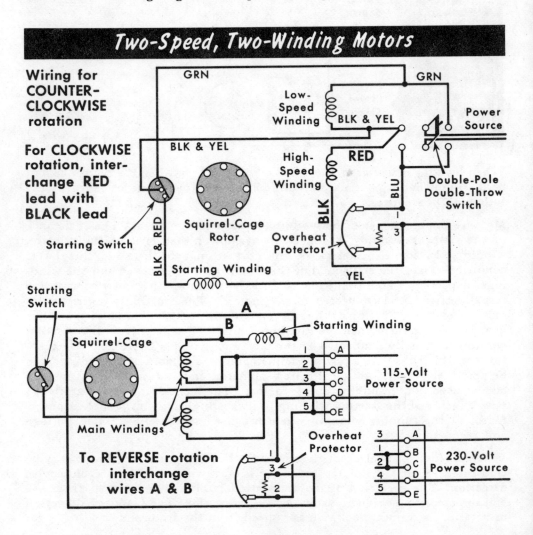

Two-Speed, Two-Winding Motors

Wiring for COUNTER-CLOCKWISE rotation

For CLOCKWISE rotation, interchange RED lead with BLACK lead

Starting Switch

GRN

Low-Speed Winding

BLK & YEL

GRN

Power Source

BLK & YEL

High-Speed Winding

RED

BLU

BLK

Double-Pole Double-Throw Switch

Squirrel-Cage Rotor

Overheat Protector

1

3

BLK & RED

Starting Winding

YEL

Starting Switch

A

B

Starting Winding

Squirrel-Cage

115-Volt Power Source

A
B
C
D
E

1
2
3
4
5

Main Windings

Overheat Protector

1
3
2

To REVERSE rotation interchange wires A & B

230-Volt Power Source

3
1
2
4
5

A
B
C
D
E

SPLIT-PHASE MOTORS

Troubleshooting Split-Phase Motors

SYMPTOM	POSSIBLE CAUSE	REMEDY
Failure to start	Blown fuse or circuit breaker open	Check rated capacity of fuses or circuit breaker; must be rated at least 125% of full-load current.
	Low voltage or no voltage	Check power source and see that voltage is within 10% of motor nameplate voltage. Low voltage causes low starting torque. The motor will not be able to start but will draw locked rotor current, causing the overheat protector to trip or the windings to burn out.
	Open circuit	Turn motor by hand. If motor rotates freely, check connections. Check starting switch. Check contacts for cleanliness and good tight contact. Check wires for breaks. If thermal overload is of the manual type, it must be reset. If all of the above check out, give shaft a good spin. If motor starts and comes up to speed (no load), the starting winding circuit is open.
	Improper line connections	Check motor connection diagram.
	Excessive load	If motor starts when disconnected from load, it may be overloaded, incorrectly applied, or driven device may be locked or bound so that it does not turn freely. Load should be reduced to correspond with the motor rating.
	Excessive end play	If motor fails to start but runs if started by hand, check the shaft for excessive end play. If this occurs, add washers on shaft journal.
	Bearings frozen	(See *Overheating of bearings*.)
Excessive noise	Unbalance	Vibration can be remedied by rebalancing rotor or pulley.
	Bent shaft	Shaft must be straightened and rotor balance checked.
	Loose parts	Check motor accessories; make sure they are properly tightened. Motor must be checked for firm mounting.
	Faulty alignment	Align motor properly with driven device.
	Worn bearings	(See *Overheating of bearings*). A dry or slightly worn bearing results in slip frequency noise or purr. Oil bearings properly (periodically as required) or replace.
	Dirt in air gap	Dismantle motor and blow out dirt with dry compressed air.
	Uneven air gap	Check for bent shaft or worn bearings. If necessary, true up rotor outside diameter or stator bore. Too great an increase in size of air gap may result in overheating.
Overheating of bearings	Motor needs oil	Oil motor with a good-grade oil as recommended by the manufacturer.
	Dirty oil	Clean out old oil and replace with a good-grade oil.

Troubleshooting Split-Phase Motors (contd.)

SYMPTOM	POSSIBLE CAUSE	REMEDY
	Oil not reaching shaft	If motor utilizes a wick, check wick and means provided for holding wick against the shaft. If motor is of the ring type, make sure the ring rotates.
	Excessive grease	If motor bearings are of the ball-bearing type, excessive grease causes them to overheat. Remove excess grease.
	Excessive belt tension	Adjust belt tension to proper value.
	Rough bearing surface	Bearing must be replaced before shaft is damaged.
	Bent shaft	(See *Excessive noise*.)
	Misalignment of shaft and bearing	(See *Faulty alignment* under *Excessive noise*.)
	Excessive end thrust	May be caused by rotor incorrectly located on shaft or by driven device causing excessive thrust on motor bearings. Check motor and application.
	Excessive side pull	Check application of motor to driven device. Make sure belt tension is not excessive.
Overheating of motor	Obstruction of ventilating system	Electric motors must be kept clean and dry, necessitating periodic inspections. Dust is removed, and ventilating openings must not be covered by other objects.
	Overloading	Check driven device to make sure it does not bind. Check application by measuring input watts and amps under normal operating conditions.
	Rotor making contact with stator	Check motor for worn bearings or bent shaft.
Rotor or stator burned out	Worn bearings	(See *overheating of bearings*.)
	Moisture	Excessive moisture can weaken insulation. Replaced parts should have special varnish treatment.
	Acids or alkalies	Special varnish treatment may be required.
	Harmful dust accumulation	Dust and dirt which are highly conductive contribute to insulation breakdown.
	Overloading	(See *Overheating*.)
	Start winding switch defective	If the rotary switch fails to disconnect starting winding from the circuit, starting winding can burn out.

QUESTIONS AND PROBLEMS

1. What makes a split-phase motor run?

2. How is starting torque obtained in a split-phase motor?

3. Name three characteristics of the running winding. Of the starting winding.

4. What occurs in the split-phase motor when the running speed approaches the synchronous speed?

5. What is the purpose and an advantage of the use of a centrifugal starting switch in the split-phase motor?

6. What is the purpose of the motor protector? Explain its operation in conjunction with the operation of the split-phase motor.

7. Name the major components of the split-phase motor.

8. How can we determine the lead connections in the split-phase motor? Which is the best method of determination?

9. What is the standard direction of rotation?

10. Describe, by means of a simple sketch, the difference between 115-volt and 230-volt operation of a two-phase motor. Explain why the connections are so made.

11. What are some of the reasons that the motor may not start?

12. What can cause the motor bearings to burn out?

Capacitor Motors — Capacitor-Start Motors

Capacitor motors are an improvement over the split-phase motors. They have become very popular for applications such as refrigerators, stokers, compressors, and air conditioners, where high starting torques are required. Capacitor motors are divided into five major classifications: capacitor-start, capacitor-start-and-run, permanent-split capacitor, two-speed, and multi-speed motors.

What makes a capacitor-type motor run? A capacitor-type motor is essentially a split-phase motor. You may recall that the split-phase motor has two windings out of phase with each other. The phase difference, or split, provides the necessary torque for starting the rotor spinning and making the spin continue. The capacitor motor improves on the split-phase motor by adding a capacitor somewhere within the motor circuit. The effect of the capacitor is to add to the motor torque, giving us a more powerful motor.

A capacitor-start motor uses a capacitor to add to the initial starting torque provided by the stator windings. Circuitwise, a capacitor motor can be made from a split-phase motor simply by connecting a capacitor in the motor circuit at the proper point. From the schematic diagrams shown, the only difference consists in the insertion of a capacitor in series with the start winding and the removal of the resistor formerly in the circuit.

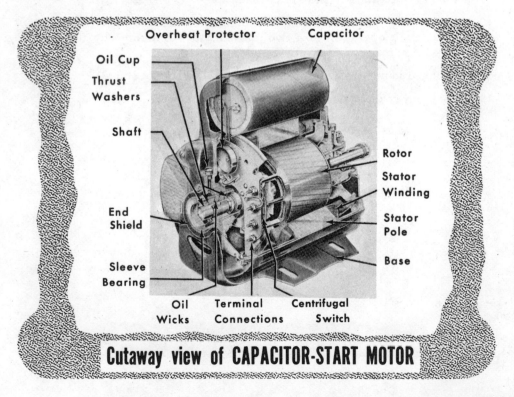

Cutaway view of CAPACITOR-START MOTOR

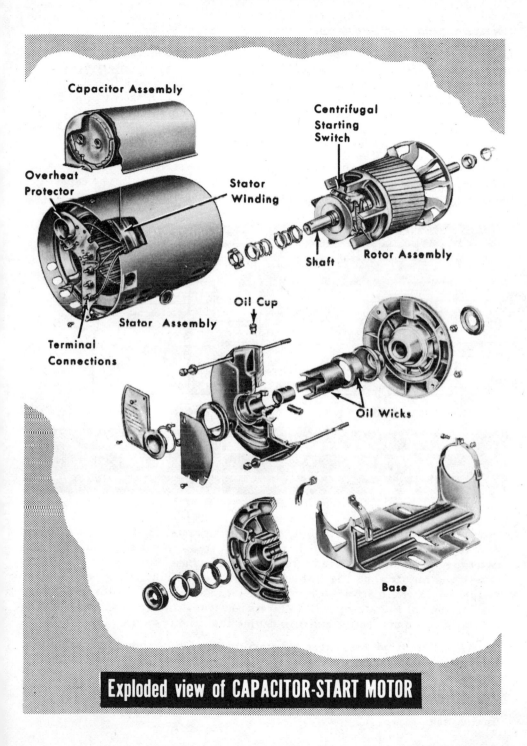

Capacitor Assembly

Centrifugal
Starting
Switch

Overheat
Protector

Stator
Winding

Shaft

Rotor Assembly

Oil Cup

Stator Assembly

Terminal
Connections

Oil Wicks

Base

Exploded view of CAPACITOR-START MOTOR

23

CAPACITOR MOTORS

Comparing the Split-Phase and the Capacitor-Start Motor

Even though these two types of motors appear to be similar, a split-phase motor cannot be converted to a capacitor motor by merely inserting a capacitor in series with its starting winding. The windings of these types of motors are designed differently. These windings have different proportions to each other, depending on the application for which they are designed. Capacitor-start motors have an inherently higher torque than the corresponding torque of a split-phase motor.

In a split-phase motor, resistance is added to the starting winding so that the current is more nearly in phase with the line voltage. In the capacitor-start motor, the capacitor causes the current in the starting winding to lead

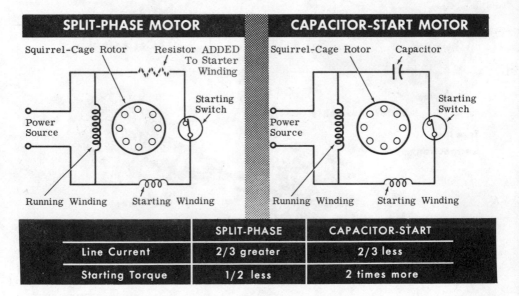

		SPLIT-PHASE	CAPACITOR-START
Line Current		2/3 greater	2/3 less
Starting Torque		1/2 less	2 times more

the running phase voltage, thereby obtaining a greater angle of displacement between the starting and the field windings. In a typical comparison of the two types of motors, the line current of a split-phase motor was found to be two-thirds higher than the line current of a corresponding capacitor-start motor, while the starting torque of the capacitor-start motor was twice that of the split-phase motor. We can see from this comparison that the capacitor is a much better starting device than the resistor.

Several different types of capacitor-start motors can be found in current use. Each type is designed differently because of the difference in required horsepower ratings and the type of capacitor used. One means of determining the type of motor is: if the motor is rated at less than 1/4 hp, it is usually single-voltage operated; if it is rated larger than 1/4 hp, it is probably a dual-voltage type.

CAPACITOR MOTORS

Capacitors Used in Capacitor-Start Motors

In the single-voltage capacitor-start motor, two types of capacitors may be used: plain paper dielectric and electrolytic. The latter type usually has a capacitance higher than one microfarad (µf) and is capable of withstanding a wider range of voltage ratings depending upon its size and the type of dielectric used. The paper capacitor, on the other hand, has a capacitance usually of the order of fractions of a microfarad (0.1 µf to 0.0047 µf) and can withstand only up to approximately 600 volts. Since one type has the advantage of large capacitance and higher voltage ratings, while the other has smaller physical size in its favor, the use of either depends on the type of operation, motor size and rating, and motor construction. The table gives some typical capacitor ratings (electrolytic) used on some of the larger horsepower capacitor-start motors.

ELECTROLYTIC CAPACITORS FOR CAPACITOR-START MOTORS

Motor Rating (hp)	Typical Capacitance and Voltage Rating (microfarads and working volts a-c)	
1/8	72/88	115
1/6	88/108	115
1/4	108/130	115
	124/149	115
1/3	161/193	115
1/2	216/259	115
3/4	378/440	115
1	378/440	115

Some of the older capacitor-start motors use a combination of transformer and capacitor instead of a straight capacitor. The advantage of the combination is that a small capacitance can be made to act as a much larger capacitance, while still retaining its small physical size. As shown, it strongly resembles the input tank circuit of a simple radio receiver. In some cases, we are able to use a small paper capacitor in conjunction with a small auto-transformer instead of a large electrolytic capacitor. However, we must remember that while the capacitance may have been increased, the capacitor working voltage remains the same. This method of increasing capacitance is rapidly becoming obsolete.

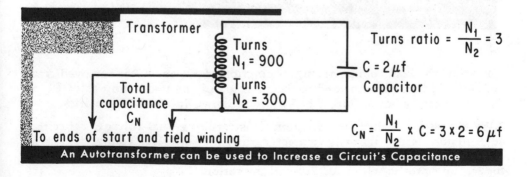

An Autotransformer can be used to Increase a Circuit's Capacitance

$$\text{Turns ratio} = \frac{N_1}{N_2} = 3$$

$$C = 2\,\mu f \quad \text{Capacitor}$$

$$C_N = \frac{N_1}{N_2} \times C = 3 \times 2 = 6\,\mu f$$

Reversible and Dual-Voltage Capacitor-Start Motors

The capacitor-start motor can be made either as a nonreversible- or a reversible-rotation motor. Some capacitor-start motors are now found for which the direction of rotation can be changed while the motors are in operation. As shown, connections are made for 115-volt operation only. For 230-volt

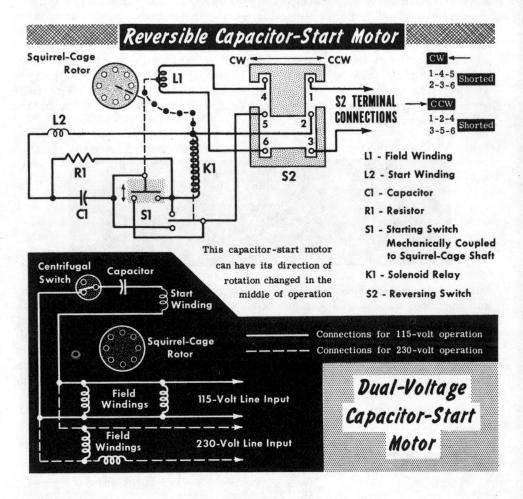

Reversible Capacitor-Start Motor

Squirrel-Cage Rotor

CW ← → CCW

L1

4   1
5   2
6   3

S2 TERMINAL CONNECTIONS

CW ←
1-4-5
2-3-6   Shorted

CCW
1-2-4
3-5-6   Shorted

L2

R1   K1

C1   S1   S2

L1 - Field Winding
L2 - Start Winding
C1 - Capacitor
R1 - Resistor
S1 - Starting Switch
        Mechanically Coupled
        to Squirrel-Cage Shaft
K1 - Solenoid Relay
S2 - Reversing Switch

This capacitor-start motor can have its direction of rotation changed in the middle of operation

Centrifugal Switch   Capacitor

Start Winding

Squirrel-Cage Rotor

———— Connections for 115-volt operation
- - - - Connections for 230-volt operation

Field Windings   115-Volt Line Input

Field Windings   230-Volt Line Input

Dual-Voltage Capacitor-Start Motor

operation, the lead coming from the junction of K1 and L2 is removed from contact 2 of S2 and connected to the junction of the two windings of L1, as shown by the dot-dash line. The switch is shown in the OFF position.

Also shown is a schematic diagram illustrating a dual-voltage, capacitor-start motor. If the motor is connected with the solid-line connections, it is wired for 115-volt operation. If instead, the dotted-line connections are made, the motor is wired for 230-volt operation.

Capacitor-Start-and-Run Motors

Basically the capacitor-start-and-run motor is very similar to the capacitor-start motor. This motor is sometimes called a two-value capacitor motor because the motor starts with one value of capacitance in series with the starting winding but runs with a different capacitance. The change of capacitance is effected by either an autotransformer or two separate capacitors.

The theory of operation of the two-value capacitor motor is based on that of the capacitor-start motor. The winding arrangement is the same for both motors. The difference between the two motors is that, in the two-value capacitor motor, a running capacitor is permanently connected in series with the starting winding and connected in parallel with the starting capacitor. At start-up, the capacitance of both capacitors are added to the circuit to obtain a high starting torque. When the motor reaches full-load speed, the centrifugal starting switch opens, and the starting capacitor is dropped out of the circuit. The running capacitor remains in the circuit, giving the motor a higher running torque. Although the remaining capacitance is reduced, there is still more torque than if there were no capacitance at all. The autotransformer-type motor operates on the same principles. The principles for changing the direction of rotation and obtaining dual-voltage operation are the same, as discussed on the preceding page.

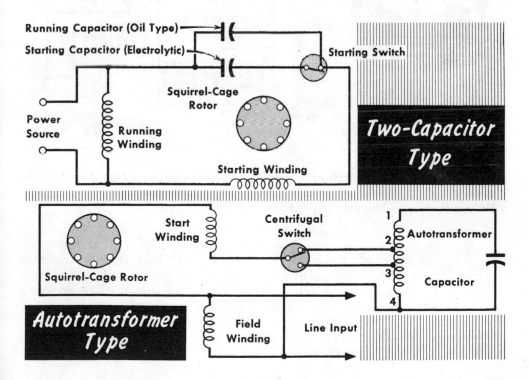

CAPACITOR MOTORS

Permanent-Split Capacitor Motors

Permanent-split capacitor motors, sometimes called single-value capacitor motors, are similar to two-phase motors connected to a single-phase line. Both the running winding and the starting winding, with its starting capacitor connected to it in series, are permanently connected to the line at all times.

The major difference between the permanent-split capacitor-type motor and the two capacitor motors just discussed is that there is no centrifugal starting switch employed. In other words, the same capacitance is used during

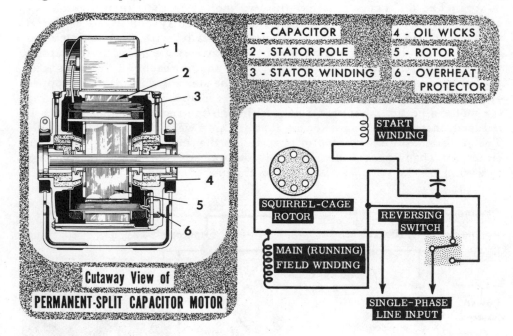

1 - CAPACITOR   4 - OIL WICKS
2 - STATOR POLE   5 - ROTOR
3 - STATOR WINDING   6 - OVERHEAT PROTECTOR

Cutaway View of
PERMANENT-SPLIT CAPACITOR MOTOR

START WINDING

SQUIRREL-CAGE ROTOR

REVERSING SWITCH

MAIN (RUNNING) FIELD WINDING

SINGLE-PHASE LINE INPUT

both the starting and running operations of the motor. As compared to the capacitor-start and capacitor-start-and-run motors, the permanent-split capacitor motor has a low starting torque.

Permanent-split capacitor motors are used primarily for fans and blowers, in oil burners, air conditioners, and similar applications where a high starting torque is not required. Reversing a permanent-split capacitor motor can be accomplished in the same manner as for the capacitor-start motor. External control of reversing action can be accomplished as shown. The reversing switch changes the capacitor connection. In one position, the capacitor is in series with the starting winding. In the other position, the capacitor is in series with the field winding. As the capacitor is switched, rotation is changed from one direction to the other.

Two-Speed and Multi-Speed Motors

Other types of motors are the two-speed and multi-speed motors. They can operate both at high and low speeds. The schematic diagram shown here illustrates the basic differences between the single- and the two-speed motor.

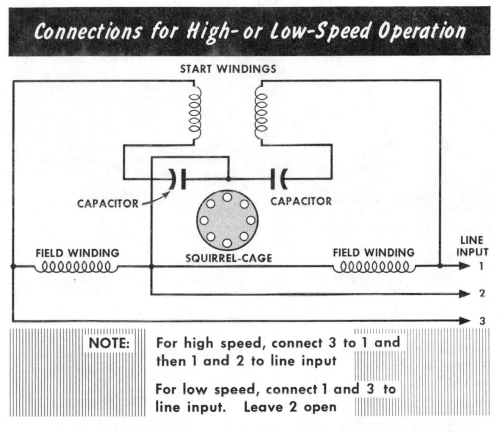

Connections for High- or Low-Speed Operation

START WINDINGS

CAPACITOR

CAPACITOR

FIELD WINDING

SQUIRREL-CAGE

FIELD WINDING

LINE INPUT

1

2

3

NOTE: For high speed, connect 3 to 1 and then 1 and 2 to line input

For low speed, connect 1 and 3 to line input. Leave 2 open

In the two-speed motor, both the starting and the field windings are in two sections, enabling both series and parallel connections of each. Motors made for this type of operation are electrically reversible only if the windings are rewound. Since this is seldom done, it is much simpler to make sure that the direction of rotation of the motor is the one required before the motor is used.

Motors of more than two speeds can also be obtained. Their circuit configuration is similar except that a multi-tapped transformer — one tap for as many speeds as required — is employed. These taps may be brought out to a multi-position rotary switch which can then be marked for the speeds desired.

CAPACITOR MOTORS

Troubleshooting Capacitor Motors

SYMPTOM	POSSIBLE CAUSE	REMEDY
Failure to start	Low or no voltage	See that voltage at power source is within 10% of motor's nameplate volts. Low voltage causes low starting torque. Motor cannot start but draws locked rotor current, causing overheat protector to trip, windings to burn out.
	Blown fuse or circuit breaker open	Check rated capacity of fuses or circuit breaker; should be rated at least 125% of full-load current.
	Open circuit	Rotate motor by hand. If it rotates freely, check connections. Check starting switch. Check contacts for cleanliness and tightness, wires for breaks. If thermal overload is manual type, reset it. If all the above check out, give shaft a good spin. If motor starts and comes up to speed (no load), starting winding circuit is open. An open capacitor has same effect as open start winding. A shorted capacitor results in low starting torque. A defective capacitor must be replaced.
	Improper line connections	Check motor connection diagram.
	Excessive load	If motor starts when disconnected from load, it may be overloaded, incorrectly applied, or driven device may be locked or bound so that it does not turn freely. Reduce load to correspond with motor rating.
	Excessive end play	If motor runs only when started by hand, check shaft. If excessive end play occurs, add washers on shaft journal.
	Bearings frozen	(See *Overheating of bearings*.)
Excessive noise	Unbalance	Vibration can be remedied by rebalancing the rotor or pulley.
	Bent shaft	Straighten shaft; check rotor balance.
	Loose parts	Check accessories for proper tightness and motor for firm mounting.
	Faulty alignment	Align motor properly with driven device.
	Worn bearings	(See *Overheating of bearings*.) A dry or slightly worn bearing results in slip frequency noise or purr. Oil bearings properly (periodically as required) or replace.
	Dirt in air gap	Dismantle motor and blow dirt out with dry compressed air.
	Uneven air gap	Check for bent shaft or worn bearings. If necessary, true up rotor's outside diameter or stator bore. If size of air gap is increased too greatly, overheating may result.
Overheating of bearings	Motor needs oil	Oil motor with a good-grade oil, as recommended by the manufacturer.

Troubleshooting Capacitor Motors (contd.)

SYMPTOM	POSSIBLE CAUSE	REMEDY
	Oil not reaching shaft	If motor has a wick, check wick and means provided for holding wick against shaft. If motor is of the ring type, be sure ring rotates.
	Excessive grease	If motor bearings are ball-bearing type, excessive grease will cause them to overheat. Remove excess grease.
	Excessive belt tension	Adjust belt tension to proper value.
	Rough bearing surface	Bearing must be replaced before shaft is damaged.
	Bent shaft	(See *Excessive noise.*)
	Misalignment of shaft and bearing	(See *Faulty alignment* under *Excessive noise.*)
	Excessive end thrust	May be caused by rotor incorrectly located on shaft or by driven device causing excessive thrust on bearings. Check motor and application.
	Excessive side pull	Check application of motor to driven device. Make sure that belt tension is not excessive.
Overheating of motor	Obstruction of ventilating system	Electric motors must be kept clean and dry. Inspect periodically. Remove dust, clear ventilating openings.
	Overloading	See that driven device does not bind. Check application by measuring input watts and amperes under normal operating conditions.
	Rotor making contact with stator	Check motor for worn bearings or bent shaft.
Rotor or stator burned out	Worn bearings	(See *Overheating of bearings.*)
	Moisture	Excessive moisture can weaken insulation. Replaced parts should have special varnish treatment.
	Acids or alkalis	Special varnish may be applied.
	Harmful dust accumulation	Highly conductive dust and dirt contribute to insulation breakdown.
	Overloading	(See *Overheating.*)
	Start winding switch defective	If rotary switch fails to disconnect start winding from circuit, start winding can burn out.

QUESTIONS AND PROBLEMS

1. What is one major advantage of a capacitor motor?

2. Name the five major classifications of capacitor-type motors.

3. What is the difference between the split-phase and capacitor-type motor? How can the capacitor motor be made from a split-phase motor?

4. If the horsepower rating of the capacitor motor is known, how do we determine the type of motor?

5. Since paper capacitors are so much smaller in size than electrolytic capacitors, why are electrolytic capacitors used almost exclusively?

6. Explain the theory behind the reversible-capacitor motor.

7. What is the basic difference between a capacitor-start motor and a capacitor-start-and-run motor? What is the major advantage of the latter over the former?

8. What is the basic principle of the permanent-split capacitor motor?

9. What is the basic difference between the permanent-split capacitor motor and the capacitor-start; and capacitor-start-and-run motors?

10. How is reversing action accomplished in the permanent-split capacitor motor?

11. Explain, by means of a simple sketch, the theory of operation of two-speed capacitor motors?

12. What are some of the reasons a capacitor-type motor may fail to start?

13. What could cause a burned out rotor or stator?

Repulsion Motors

Repulsion-type motors are no newcomers to the motor field. Various types of repulsion motors can be found, each using the repulsion principle for starting. Motors included under the repulsion heading are: the straight repulsion motor, which starts and runs on the magnetic repulsion principle;

REPULSION-INDUCTION MOTOR PARTS

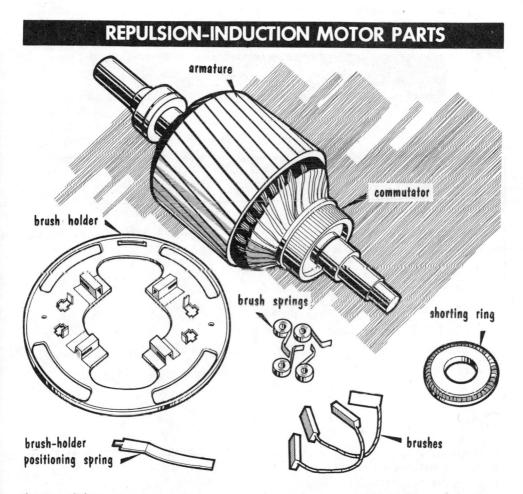

armature

commutator

brush holder

brush springs

shorting ring

brush-holder positioning spring

brushes

the repulsion-start induction motor, which starts on the repulsion principle but at a predetermined speed operates as a single-phase induction motor; the repulsion-induction motor, which uses the repulsion-start principle but has an additional armature winding that provides additional running torque; and the electrically reversible motor, which is either a straight repulsion or repulsion-induction motor specifically connected for reversing action. The parts of a typical repulsion motor, excluding the frame and stator, are illustrated above.

Repulsion Motors (contd.)

What makes the repulsion-type motor run? We know that both a fixed winding (stator) and a rotating winding (rotor or armature) are needed to make up a motor. We know also that the stator is laminated and the field windings are wound around protrusions called field poles. Current flowing through these windings creates a magnetic field around the poles. The number of such poles within the motor determines its running speed. We now

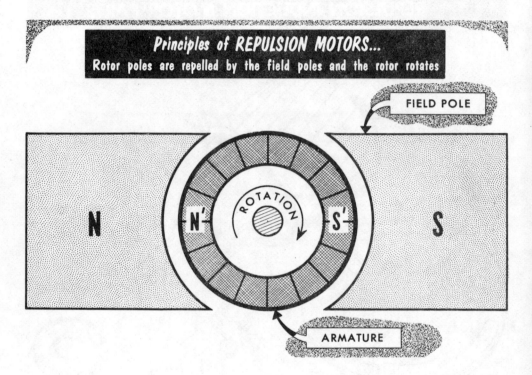

Principles of REPULSION MOTORS...

Rotor poles are repelled by the field poles and the rotor rotates

FIELD POLE

N   N'   ROTATION   S'   S

ARMATURE

insert a rotatable winding within the stator and mount it so that it is free to rotate within the stator. With the a-c voltage applied to the field winding, an a-c magnetic field is set up. The rotor is wound like a universal motor's armature. The brushes are tied together. The only current in this rotor is induced by the contracting and expanding field of the stator. Because the brushes furnish the only complete circuit for the current, the poles of the magnetic field induced in the rotor are located opposite the brushes. The magnetic fields of the stator and the rotor align as shown. Since like poles repel each other, we can assume that the rotor poles are repelled by the field poles, and our rotor rotates. This is the operating principle of the repulsion motor.

Repulsion Motors (contd.)

The straight repulsion motor is a single-phase motor, often called an inductive series motor. It starts and runs on the induction principle and is a varying-speed type of motor. Its speed and torque characteristics are similar to that of a d-c series motor.

The field windings are series-connected to the line input, but not connected in any way to the rotor. The motor brushes are tied together within the motor and so situated that their axis is inclined at a very small angle to the neutral axis of the field. A diagram of the straight repulsion motor is shown here.

Suppose that the field poles are N and S and that lines of flux are passing between them. Then the neutral axis of this flux field is the line OO'

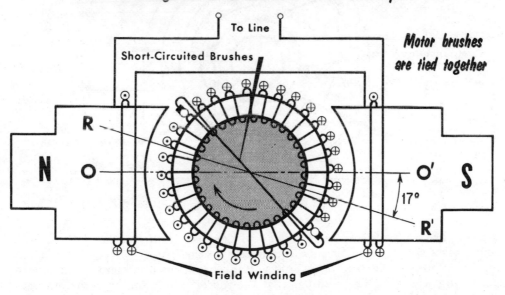

drawn through the center of the field poles. If this is the case, then the placing of the brushes is incorrect. However, they have been drawn in this position for illustration only. If the brushes were shown as they should actually be placed, their centers would lie along the line RR', at an angle of approximately 17° from line OO'. At this angle maximum starting torque is obtained. Better commutation and longer life are possible if the angle is made slightly greater than 17°.

Compensated Repulsion Motors

Some straight repulsion motors have an additional winding in the stator which acts as a compensating winding. Compensating windings are discussed in greater detail in the chapter covering universal motors, but for clarity will be reviewed here.

A compensating winding is used to neutralize any cross-magnetization set up in the armature by the flux passing through it. In the repulsion motor, this winding is connected to an extra insulated set of brushes. These

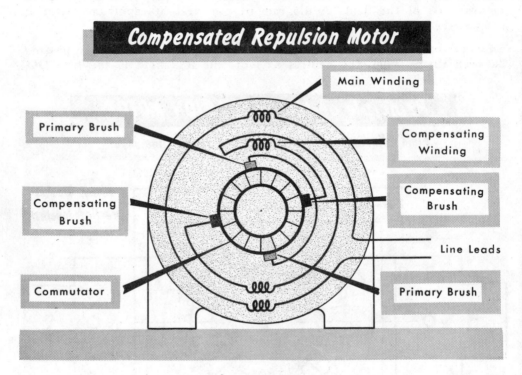

brushes are not connected in any way to the rest of the motor. When current passes through the main field winding on the motor, voltage is induced in the compensating winding. This voltage produces a current in the compensating winding. The result is that the motor power factor is maintained at close to unity, and the speed of rotation is almost constant.

Straight repulsion motors are constructed so that any number of brushes may be used, depending on the type of field winding and the number of poles in the field winding. For instance, if the field is lap wound, there must be one brush for each field pole. However, if the field is lap wound and cross-connected, only two brushes are necessary, regardless of the number of field poles. A more detailed description follows, based on a four-pole motor.

A Four-Pole Repulsion Motor

In some repulsion motors, the poles are wound concentrically; in others, progressively. In our case, the motor is wound concentrically. The brushes are usually mounted on a ring which can be rotated in either a counterclockwise or clockwise direction. If this ring is shifted so that the brushes are on the lines AA and CC, the armature produces zero torque. This is called the torque neutral or hard neutral point on the motor. Similarly, if the brushes are set on the lines BB and DD, the armature again produces zero torque. This point is the soft neutral. For maximum starting torque, the brushes must be set on the line which is shown as being shifted 17 electrical degrees from line AA. If a mistake is made, and the brush shift is made from the soft neutral lines BB and DD, the motor does not operate properly. To check whether this condition exists, simply rotate the brush ring in either direction. If the armature rotates in the direction of shift, the setting is correctly made from the hard neutral; otherwise, the brushes must be reset from the hard neutral.

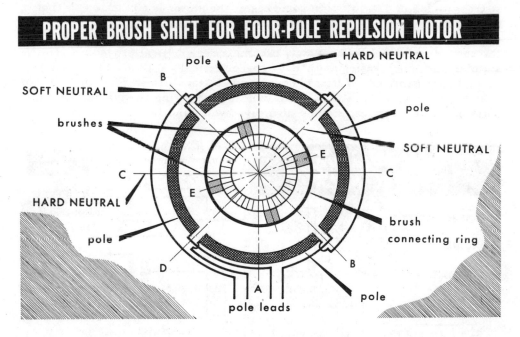

Our illustration shows the motor with four brushes. Depending on the type of field winding, the straight repulsion motor can work with more or less than four brushes. If the motor has wave or cross-connected windings, only two brushes are required; they must be set 180 electrical degrees apart on the commutator. If lap windings are used, for every field pole in the motor, there must be an accompanying brush.

REPULSION MOTORS

A Four-Pole Repulsion Motor (contd.)

Determining the type of winding used in the motor is a simple procedure. If removal of any two opposing brushes causes motor stoppage, the winding is either a wave type or a lap type cross-connected. If removal of any one brush causes motor stoppage, the winding is a lap type non-cross-connected.

If the brushes are set from the hard neutral, any brush shift causes the armature to rotate in the direction of shift from the hard neutral. Thus, we can reverse direction of rotation, provided we know where the hard neutral is located. The procedure for locating the hard neutral is presented in the section on care and maintenance of motors. Once the hard neutral has been located, reversing the direction of rotation is simple. Simply shift the brushes away from the hard neutral point in the direction desired. Care must be taken not to shift them too far, especially in multi-pole motors.

The straight repulsion motor can be operated from two different voltage sources — 115 or 230 volts. In some cases, the motor is connected so that it may be made a variable-speed type, either through mechanical or electrical methods. The mechanical method involves increased brush shift, which, in the case of multi-pole motors, may become a very critical procedure. In the simpler and much easier method — the electrical — a variable resistor, a variable transformer (Variac), or a tapped transformer may be used to change the input voltage to the field winding, changing the no-load speed of rotation. The electrical method also saves wear and tear on brushes.

DUAL-VOLTAGE CONNECTIONS

FIELD WINDING
BRUSH
COMMUTATOR
TERMINAL BLOCK
110 VOLTS
220 VOLTS

TOP TERMINAL SHOWS CONNECTION TO MOTOR WINDINGS ONLY. FOR PROPER CONNECTION TO A-C CIRCUIT, CHECK VOLTAGE SOURCE AND CONNECT AS INDICATED ON LOWER TERMINALS.

NOTE: Dotted lines show possible connections

VARIABLE RESISTOR
115 VOLTS OR VARIABLE TRANSFORMER
115 VOLTS
115 VOLTS OR TAPPED TRANSFORMER
TERMINAL BLOCK
MOTOR FIELDS

VARIABLE-SPEED CONNECTIONS

The Repulsion-Start Induction Motor

The repulsion-start induction motor differs from the straight repulsion motor in one respect. The starting principles are the same, but, at a predetermined speed, a special device is actuated which short-circuits all the commutator windings. From this point on, the motor operates as a single-phase induction motor.

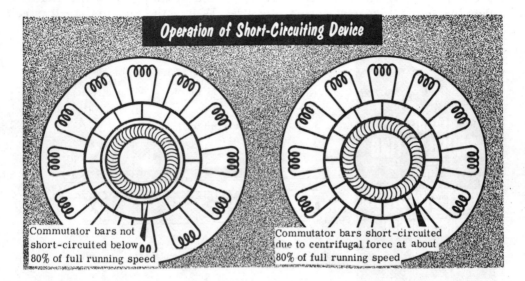

Operation of Short-Circuiting Device

Commutator bars not short-circuited below 80% of full running speed

Commutator bars short-circuited due to centrifugal force at about 80% of full running speed

The principle behind this special device, called the *short-circuiter*, is simple. The actuating force is centrifugal, as it is for the governor on the universal motor. As the motor starts, the torque is very high. However, as the speed of the motor increases, the torque decreases in an almost linear fashion. At approximately 80% of synchronous speed, the short-circuiter operates centrifugally, thus short-circuiting all the commutator segments and, effectively, cutting off all current flow through the brushes. The motor now operates as an induction motor. Since an induction motor functions on the magnetic-induction principle, where fields of both the stator and armature rotate in the same direction, the higher the speed at which the short-circuiter operates, the less line current is drawn by the motor.

There are two major types of repulsion-start induction motors: brush-riding and brush-lifting. The brush-riding type earns its name from the fact that after the short-circuiter operates, the brushes maintain contact with the commutator, although no current flows through them. The brush-lifing type is so called because the motor contains a mechanism which lifts the brushes off the commutator immediately after the short-circuiter is actuated. Since the latter type is the more complicated, we shall go into greater detail on the next page.

Brush-Lifting Type of Repulsion-Start Induction Motors

The usual type of commutator used for brush-lifting is the radial type, in which the commutator segments are perpendicular to the armature shaft. Several methods are used to lift the brushes off the commutator. The basic principles are the same, but the location of the components which perform the brush-lifting and short-circuiting is different. The two actions are both accomplished by the centrifugal force of the spinning armature causing weights to move against spring pressure. Where the weights are located

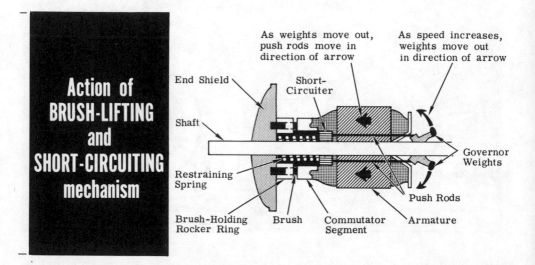

Action of BRUSH-LIFTING and SHORT-CIRCUITING mechanism

As weights move out, push rods move in direction of arrow

As speed increases, weights move out in direction of arrow

End Shield

Short-Circuiter

Shaft

Governor Weights

Restraining Spring

Push Rods

Brush-Holding Rocker Ring

Brush

Commutator Segment

Armature

at the end of the armature opposite the commutator, a nonadjustable coil spring provides the opposing pressure. As the speed of the motor increases, the weights begin to move out under the centrifugal spinning force. The weights, in turn, push against rods attached to a metal ring (short-circuiter) whose outer diameter is almost exactly the same as the inner diameter of the commutator. As the ring touches the inner sides of the commutator segments, the segments are short-circuited. However, since the motor is still speeding up, the weights continue to push against the rods and eventually push the brush-holding rocker ring away from the commutator.

Where the weights are located within the commutator, the short-circuiting and brush-lifting principle is the same as just described. However, this type motor uses a single, adjustable spring placed over the armature. Tension is set by means of an adjusting nut, a distinct advantage over the nonadjustable springs previously mentioned. However, it has the disadvantage of possible maladjustment; too much spring tension may cause operation of the short-circuiter above the motor's synchronous speed; too little spring tension can result in failure of the motor to come up to full-load speed.

The Repulsion-Induction Motor

The repulsion-induction motor is very similar to both the straight repulsion motor and the repulsion-start induction motor. It has the same starting principle as the latter two motors, but no short-circuiter is included in its construction. Instead, it combines both a repulsion and squirrel-cage winding in its armature. Both windings are always in operation while the armature rotates. The armature is so constructed that the squirrel-cage winding lies in slots which are located beneath the slots of the repulsion winding, as shown below.

At start, no torque is produced by the squirrel-cage motor. All starting torque is provided by the repulsion winding. However, once the armature begins its rotation, the voltage induced in the squirrel-cage winding produces some torque within its winding. This torque adds to the original repulsion winding torque. However, after the motor reaches approximately one-half synchronous speed, the combined torque begins to decrease with the increasing speed. If the motor reaches synchronous speed, the squirrel-cage continues to add torque to the repulsion winding, as opposed to the squirrel cage of a simple induction motor. However, if we exceed the synchronous

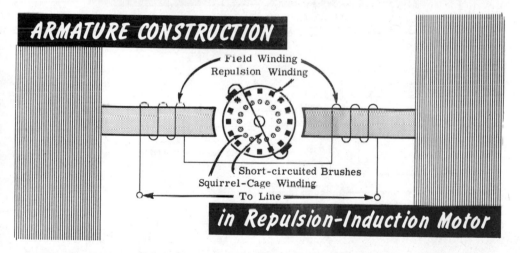

ARMATURE CONSTRUCTION

Field Winding
Repulsion Winding

Short-circuited Brushes
Squirrel-Cage Winding
To Line

in Repulsion-Induction Motor

speed more than slightly, the squirrel-cage winding produces a torque in opposition to that of the repulsion winding and effectively becomes a braking force within the motor.

The characteristics of the repulsion-induction motor are very similar to that of the straight repulsion type. Direction of rotation and setting of the neutral are the same. In fact, it is sometimes difficult to differentiate between the two. An easy way to determine which is which is to remove the load, start the motor and then remove the brushes. If the motor continues to run, it is a repulsion-induction type.

Reversible Repulsion Motor

The repulsion motor can be made electrically reversible. The other repulsion motors which we discussed operate with either 115 or 230 volts. The electrically reversible motor can be operated from either source, depending upon construction, but not from both. All three of the discussed types of repulsion motors can be operated from either source, depending upon construction, but not from both. All of the types of repulsion motors discussed can be made for reversing operation. A special construction of the field winding makes it possible. Reversal is accomplished simply by changing the reversing switch's position.

The brushes of reversible repulsion motors must be set exactly on the hard neutral with no brush shift at all. The procedure for setting the hard neutral is exacting and must be done very carefully as follows: Connect Phase B winding to one side of the line input. Connect Phase C winding to the

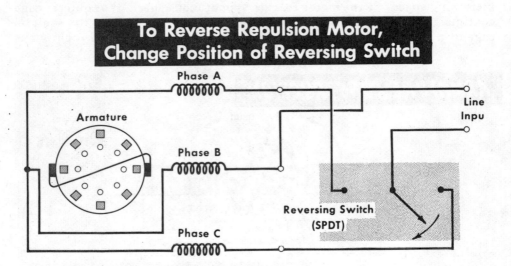

remaining side of the line input. Refer to the discussion for setting the hard neutral for the repulsion motor in the section on care and maintenance of brushes and set the hard neutral in the same manner. Note the brush setting. Remove Phase C from the line and connect Phase A to the line. Repeat the hard neutral setting procedure, and again note the brush setting. Set the brushes at the point between the two settings obtained previously. From the illustration, it can be seen that the brushes are directly in line with Phase B. If this is not so, the motor has unequal torques in different directions of rotation. If the electrically reversible repulsion motor contains a short-circuiter and brush-lifting device, their operation is identical to that used in the nonreversible types. Refer to the previous discussion for explanation of the short-circuiting and brush-lifting mechanisms.

REPULSION MOTORS

Troubleshooting Repulsion-Type Motors

SYMPTOM	POSSIBLE CAUSE	REMEDY
Failure to start	Blown fuse or circuit breaker open	Check rated capacity of fuses or circuit breaker; must be rated at least 125% of full-load current.
	Low voltage or no voltage	Check power source and see that voltage is within 10% of motor nameplate voltage.
	Open circuit	Rotate motor by hand. If motor rotates freely, check connections. Check contacts for cleanliness and for tightness, wires for breaks. An open circuit may be indicated by excessive sparking at the brushes. See that starting switch, if one is used, is closed. If a speed control governor is used, see that it is closed when motor is at rest.
	Improper line connections	Check motor connection diagram.
	Excessive load	If motor starts when disconnected from load, it may be overloaded, incorrectly applied, or driven device may be locked or bound so that it does not turn freely. Reduce load to correspond with motor rating.
	Brushes worn or sticking	Brushes may be worn so that only very light, or no, contact is made with commutator. Worn brushes must be replaced. If brushes are sticking in holders, brush springs may be weak, or commutator dirty. Another possibility may be high mica. Clean commutator with fine sandpaper, *never with emery!*
	Brushes incorrectly set	Check that brushes are set the proper distance off neutral.
	Excessive end play	If motor fails to start, but runs if started by hand, check shaft for excessive end play. If this occurs, add washers on shaft journal.
	Bearings frozen	(See *Overheating of bearings.*)
Excessive noise	Unbalance	Vibration can be remedied by rebalancing the rotor or pulley.
	Bent shaft	Shaft must be straightened and rotor balance checked.
	Loose parts	Check motor accessories for proper tightness and motor for firm mounting.
	Faulty alignment	Align motor properly with driven device.
	Worn bearings	(See *Overheating of bearings.*) A dry or slightly worn bearing results in slip frequency noise or purr. Oil bearings properly (periodically as required) or replace.
	Dirt in air gap	Dismantle motor and blow dirt out with dry compressed air.
	Uneven air gap	Check for bent shaft or worn bearings. If necessary, true up outside diameter of rotor or stator bore. Too great an increase in size of air gap may result in overheating.

REPULSION MOTORS

Troubleshooting Repulsion-Type Motors (contd.)

SYMPTOM	POSSIBLE CAUSE	REMEDY
Overheating of bearings	Motor needs oil	Oil motor with good-grade oil as recommended by manufacturer.
	Dirty oil	Clean out old oil and replace with a good grade of oil.
	Oil not reaching shaft	If motor utilizes a wick, check wick and means provided for holding wick against shaft. If motor is of the ring type, make sure the ring rotates.
	Excessive grease	If motor bearings are of ball-bearing type, excessive grease causes them to overheat. Excess grease must be removed.
	Excessive belt tension	Adjust belt tension to proper value.
	Rough bearing surface	Bearing must be replaced before shaft is damaged.
	Bent shaft	(See *Excessive noise.*)
	Misalignment of shaft and bearing	(See *Faulty alignment* under *Excessive noise.*)
	Excessive end thrust	May be caused by rotor incorrectly located on shaft or by driven device causing excessive thrust on motor bearings. Check motor and application.
	Excessive side pull	Check application of motor to driven device. Make sure the belt tension is not excessive.
Excessive brush wear	Dirty commutator	Clean with fine sandpaper, *never with emery!*
	Improper contact with commutator	May be caused by weak brush springs or by high mica.
	Excessive load	(See *Failure to start.*)
	Governor not acting promptly	(See *Governor operating improperly.*)
	High mica	True up commutator face with a light cut.
	Rough commutator	True up commutator face with a light cut.
Overheating of motor	Obstruction of ventilating system	Electric motors must be kept clean and dry. This means periodic inspections. Dust must be removed, and ventilating openings must not be covered.
	Overloading	Check driven device to make sure it does not bind. Check application by measuring input watts and amps under normal operating conditions.
Commutator or stator burned out	Worn bearings	(See *Overheating of bearings.*)
	Moisture	Excessive moisture can weaken insulation. Replaced parts must have special varnish treatment.
	Acids or alkalies	Special varnish treatment may be required.

Troubleshooting Repulsion-Type Motors (contd.)

SYMPTOM	POSSIBLE CAUSE	REMEDY
	Harmful dust accumulation	Dust and dirt which are highly conductive contribute to insulation breakdown.
	Overloading	(See *Overloading* under *Overheating*.)
Governor operates improperly	Dirty commutator	Clean with fine sandpaper, *never with emery!*
	Governor mechanism sticking	Work governor by hand. Replace parts if necessary.
	Worn or sticking brushes	(See *Failure to start*.)
	Low frequency in supply circuit	Check frequency of supply circuit.
	Low voltage	(See *Low or no voltage* under *Failure to start*.)
	Incorrect connections	Check connection diagram for motor.
	Incorrect brush setting	Check that brushes are set the proper distance off neutral.
	Excessive load	(See *Excessive load* under *Failure to start*.)
	Incorrect spring tension	Reset spring tension or replace springs.

REPULSION MOTORS

QUESTIONS AND PROBLEMS

1. Name four types of repulsion motors.

2. What is the operating principle of a repulsion motor? Explain fully.

3. What is a straight repulsion motor and the principle upon which it starts and runs?

4. How is the straight repulsion motor constructed, and why is it so constructed? Why is the setting of the brushes so important?

5. Explain the reason behind the use of a compensating winding in the straight repulsion motor.

6. What is the difference between the soft and hard neutral points in the repulsion motor?

7. If the motor is wave wound, how many brushes are required, and how must they be placed? How many brushes are required in the lap wound repulsion motor?

8. How do we determine whether the repulsion motor is lap wound or wave wound?

9. Explain the methods by which the straight repulsion motor can be made both reversible and variable speed.

10. What is the difference between the straight repulsion motor and the repulsion-start induction motor?

11. What is the principle behind the short-circuiter?

12. What is brush-riding? Brush-lifting? On what type of commutator construction is each used?

13. Explain the operation of the short-circuiting device in conjunction with the brush-lifter.

14. Name the similarities and differences between the straight repulsion, repulsion-start-induction, and the repulsion-induction types of motors.

15. How is starting torque obtained in the repulsion-induction motor?

16. How can we differentiate between the straight repulsion and repulsion-induction motors?

17. How is the repulsion motor made reversible?

18. Describe the procedure for setting the brushes on the hard neutral point for the reversible motor.

19. Explain some of the reasons why a repulsion motor fails to start.

20. What could cause rotor or stator burnout?

Shaded-Pole Motors

A not too widely used type of motor employs a unique method of starting, called the *shading-coil* method. Motors which use this means of obtaining starting torque are called *shaded-pole* motors. Use of this type motor is limited because of its very low starting torque and high-slip operation. Their chief use is in small fans, such as desk fans and rubber-bladed automobile fans, and for fan motors in small air-conditioning units.

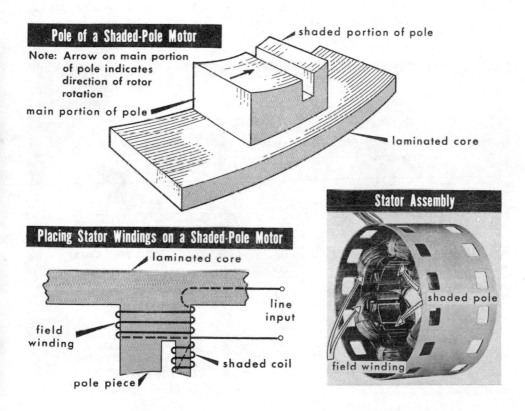

What makes the shaded-pole motor run? It requires both a stator and a rotor. However, the usual construction is not found here. True, the stator poles are made of laminated steel pieces, but their pole pieces do not resemble any other pole pieces. The shaded pole is a form of induction motor, but its windings are distributed differently than those of any other induction motor.

Each pole piece comes with a slot cut into its face. Within this slot is placed the shading coil, which can be of single- or multi-stranded wire. The chief requirement is that it form a closed circuit or loop. The main field windings are wound around the remainder of each pole piece.

Operation of the Shaded-Pole Motor

When the main field coils are energized, a magnetic field is set up between the pole pieces and the rotor. A portion of the magnetic field is also cut by the shading coil, which lies directly in the path of the magnetic field between the pole piece and rotor. The effect of the shading coil is to make the lines of flux which it cuts slightly out of phase with the remainder of the flux lines coming from the pole piece. In effect, we now have a two-phase magnetic field, similar to a two-phase voltage, at every pole piece. Provided the load is very small, this phase-shifted magnetic field has sufficient torque to start small motors with rotors of squirrel-cage construction.

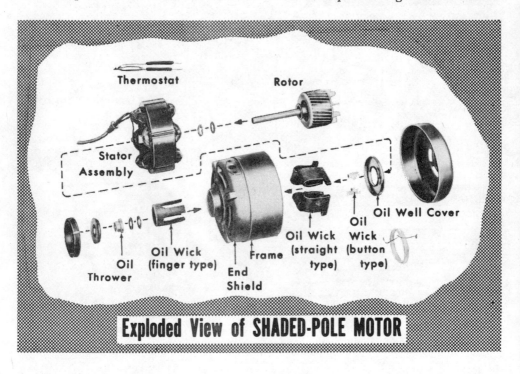

Exploded View of SHADED-POLE MOTOR

Because of its low starting torque, the shaded-pole motor uses a high-resistance squirrel-cage-type rotor. The high resistance of the rotor adds somewhat more starting torque but, in turn, causes the motor to have a large amount of slip. Because of this high slip, the motor speed is not constant and can vary over a wide range with a change in either the load or the applied voltage.

Since the shading coil is not connected to the line input, only the stator winding receives the a-c input. With only a one-winding connection, only a single-phase voltage can be applied to the motor. These motors are available for operation with 115 or 230 volts but not as dual-voltage motors.

Reversing the Shaded-Pole Motor

Mechanically, the direction of rotation of a shaded-pole motor can be reversed by disassembling the motor and reversing the relative positions of the rotor and stator, or new slots can be cut into the pole pieces on the side opposite the original slots and the shading coils laid in these slots. Since the rotor always rotates toward the side of the pole piece in which the shading coil lies, this procedure effectively changes the direction of rotation.

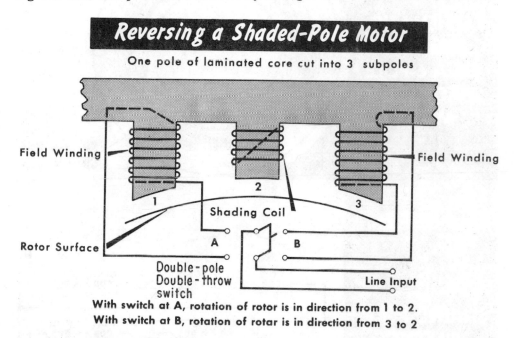

Reversing a Shaded-Pole Motor

One pole of laminated core cut into 3 subpoles

Field Winding

Field Winding

1

2

3

Shading Coil

Rotor Surface

A

B

Double-pole
Double-throw
switch

Line Input

With switch at A, rotation of rotor is in direction from 1 to 2.
With switch at B, rotation of rotar is in direction from 3 to 2

Electrically, the motor can be reversed by two means. The first method is to use two sets of field coils with each shading coil, and a switch can open or close the circuit to either set of field coils at will. The two sets of field coils are placed upon two outer poles in a set of three poles, and the shading coil is placed on the center pole. Depending on the setting of the switch, the shading coil is either in front of the energized field coil and behind an unenergized field coil or behind an energized field coil and in front of an unenergized field coil. We know from this discussion that the rotor always rotates toward the shaded-coil portion of the field pole. Thus, switching can reverse the direction of rotation by changing the relative positions of field and shading coils.

The second method is to mount two shaded-pole motors within the same housing, connecting one for one direction of rotation and the other for the opposite direction.

Varying the Speed of a Shaded-Pole Motor

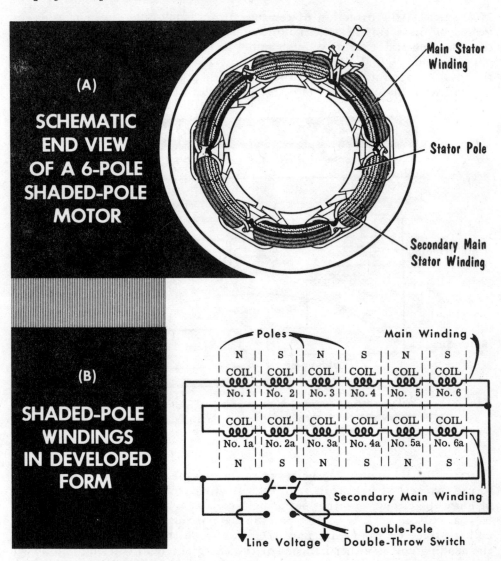

(A)

SCHEMATIC END VIEW OF A 6-POLE SHADED-POLE MOTOR

Main Stator Winding

Stator Pole

Secondary Main Stator Winding

(B)

SHADED-POLE WINDINGS IN DEVELOPED FORM

Poles — Main Winding

N S N S N S

COIL No. 1 | COIL No. 2 | COIL No. 3 | COIL No. 4 | COIL No. 5 | COIL No. 6

COIL No. 1a | COIL No. 2a | COIL No. 3a | COIL No. 4a | COIL No. 5a | COIL No. 6a

N S N S N S

Secondary Main Winding

Line Voltage

Double-Pole Double-Throw Switch

The shaded-pole motor can be made into a varying-speed motor by using a tapped transformer; or the motor can also be wound for two or more speeds. For two speeds, this is usually accomplished by placing two windings of different potential in the same slots. The figure (A) is a schematic end view of a typical six-pole, two-speed wound motor of this type. The circuit diagram shows these windings in developed form, indicating the relationship of the coils and pole pieces of the motor. The choice of speed is selected with the double-pole, double-throw switch.

Varying the Speed of a Shaded-Pole Motor (contd.)

A recent development is a shaded-pole motor wound for two-speed operation without the use of multiple windings. This is known as the "consequential-pole" motor (Patent No. 2,900,588 to L. A. Ramer of RMR Corp.).

This is accomplished by all coils being of a single winding type and located in separate slots in the same manner as for single-speed motors, the difference being that five of the six windings are of same gage wire, and the sixth winding consists of a smaller gage wire and has more turns than the other five windings. The pole of this sixth winding is the consequential pole. The circuit diagram shows these windings in developed form.

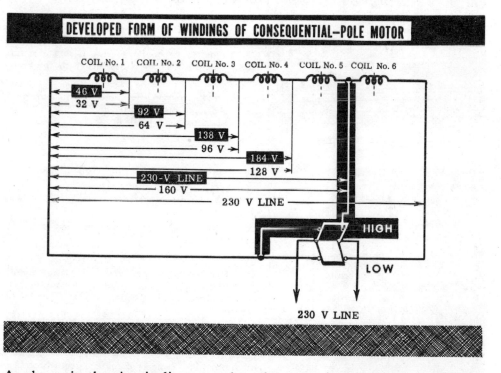

DEVELOPED FORM OF WINDINGS OF CONSEQUENTIAL–POLE MOTOR

As shown in the circuit diagram, when the motor is running on low speed, all six windings are in the circuit with full line voltage across the six windings. The distribution across the windings of the line voltage is shown.

When the motor is running on high speed, full line voltage is now distributed across only five of the windings, with the sixth winding (the consequential pole) being dropped from the circuit. This decreases the impedance of the motor's winding, thereby making the motor stronger and giving a higher operating speed. The voltages shown in the illustration are for a 230-volt line source. If the line source is 115 volts, the corresponding voltages would also be cut in half.

SHADED-POLE MOTORS

Troubleshooting Shaded-Pole Motors

SYMPTOM	POSSIBLE CAUSE	REMEDY
Failure to start	Blown fuse or circuit breaker open	Check capacity of fuses or circuit breaker; should be rated at least 125% of full-load current.
	Low or no voltage	Check power source; see that voltage is within 10% of motor nameplate voltage.
	Open circuit	Check connections. Check contacts for cleanliness and tightness, and wires for breaks. If motor is variable speed, check speed control.
	Improper line connections	Check motor connection diagram.
	Excessive load	If motor starts when disconnected from load, it may be overloaded, incorrectly applied, or driven device may be locked or bound so that it does not turn freely. Reduce load to correspond with the motor rating.
	Excessive end play	If motor fails to start, but runs if started by hand, check the shaft for excessive end play; if it occurs, add washers on shaft journal.
	Bearings frozen	(See *Overheating of bearings*.)
Excessive noise	Unbalance	Vibration can be remedied by rebalancing the rotor or pulley.
	Bent shaft	Shaft should be straightened and rotor balance checked.
	Loose parts	Check accessories for proper tightness and motor for firm mounting.
	Faulty alignment	Align motor properly with driven device.
	Worn bearings	(See *Overheating of bearings*.) A dry or slightly worn bearing results in slip frequency noise or purr. Oil bearings properly (periodically as required) or replace.
	Dirt in air gap	Dismantle motor and blow out dirt with dry compressed air.
	Uneven air gap	Check for bent shaft or worn bearings. If necessary, true up outside diameter of rotor or stator bore. Too great an increase in size of air gap may result in overheating.
Overheating of bearings	Motor needs oil	Oil with good-grade oil, as recommended by manufacturer.
	Dirty oil	Clean out old oil; replace with a good grade of oil as recommended by the manufacturer.
	Oil not reaching shaft	If motor utilizes a wick, check wick and means for holding wick against shaft. If motor is of the ring type, make sure ring rotates.

Troubleshooting Shaded-Pole Motors (contd.)

SYMPTOM	POSSIBLE CAUSE	REMEDY
	Excessive grease	If motor bearings are ball-bearing type, excessive grease causes them to overheat. Remove excess grease.
	Rough bearing surface	Bearing must be replaced before shaft is damaged.
	Bent shaft	(See *Excessive noise — Bent Shaft.*)
	Misalignment of shaft and bearing	(See *Faulty alignment* under *Excessive noise.*)
	Excessive end thrust	Rotor may be incorrectly located on shaft; driven device may be causing excessive thrust on motor bearings. Check motor and application.
	Excessive side pull	Check application of motor to driven device.
Overheating of motor	Obstruction of ventilating system	Electric motors must be kept clean and dry. Inspect them periodically. Remove dust and clear ventilating openings of other objects.
	Overloading	Make sure driven device does not bind. Check application by measuring input watts and amperes under normal operating conditions.
Rotor or stator burned out	Worn bearings	(See *Overheating of bearings.*)
	Moisture	Excessive moisture can weaken insulation. Replaced parts should have special varnish treatment.
	Acids or alkalis	Special varnish may be applied.
	Harmful dust accumulation	Dust and dirt which are highly conductive contribute to insulation breakdown.
	Overloading	(See *Overheating — Overloading.*)
Motor runs improperly	Low frequency in supply circuit	Check frequency of supply circuit.
	Low voltage	(See *Low or no voltage* under *Failure to start.*)
	Incorrect connections	Check connection diagram for motor.
	Excessive load	(See *Excessive load* under *Failure to start.*)

SHADED-POLE MOTORS

QUESTIONS AND PROBLEMS

1. Describe the stator construction of a shaded-pole motor.

2. Explain the theory of operation of a shaded-pole motor.

3. What is a major disadvantage of a shaded-pole motor?

4. List some of the most important applications of a shaded-pole motor.

5. Can the shaded-pole motor be operated as a dual-voltage motor? Explain why or why not.

6. Describe the mechanical method for reversing the direction of rotation of a shaded-pole motor. Explain the theory behind this method.

7. Explain two electrical methods for reversing the direction of rotation of a shaded-pole motor.

8. Name two methods by which a shaded-pole motor can be made to have two or more speeds.

9. Explain the theory of operation of a "consequential-pole" type of shaded-pole motor.

10. Explain some of the reasons why a shaded-pole motor might fail to start.

11. What are some of the reasons why bearings may become overheated?

12. What could cause stator or rotor burnout?

Universal Motors

Universal motors are an important development in the electric motor field. They are small, series-wound motors which can be operated from either an a-c or d-c power source. The motor performance is the same in either application, hence the term *universal*.

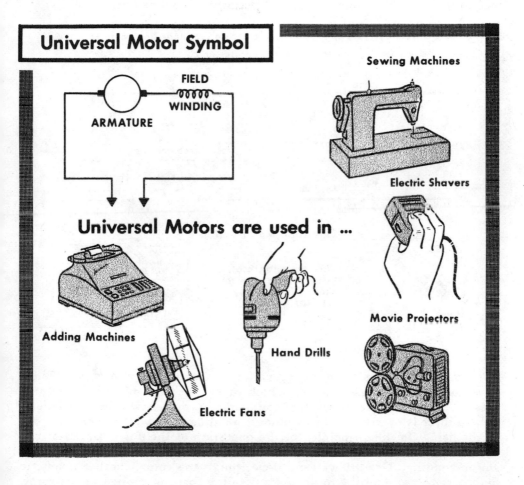

Universal Motor Symbol

FIELD WINDING

ARMATURE

Universal Motors are used in ...

Sewing Machines

Electric Shavers

Movie Projectors

Adding Machines

Hand Drills

Electric Fans

Since the universal motors are series-wound, their performance character-istics are very similar to those of a series-wound d-c motor. Universal motors are usually made in sizes from 0 to 3/4 hp. The very low fractional horsepower universal motors, which range from 0 up to 1/150 hp, are used in such equipment as sewing machines, fans, portable hand tools, motion-picture projectors, and electric shavers. The higher ratings, 1/150 to 3/4 hp, are found in vacuum cleaners, electric typewriters, motion-picture projec-tors, cameras, adding machines, and calculating machines.

Principles of Operation

What makes the universal motor run? It operates from either a single-phase a-c or a d-c source. The a-c operation is simple. The stator windings are so arranged so that, as the current alternates, the magnetic poles change from north to south, south to north, etc. By changing the polarity of the magnetic fields acting on the rotor, the rotor is caused to turn.

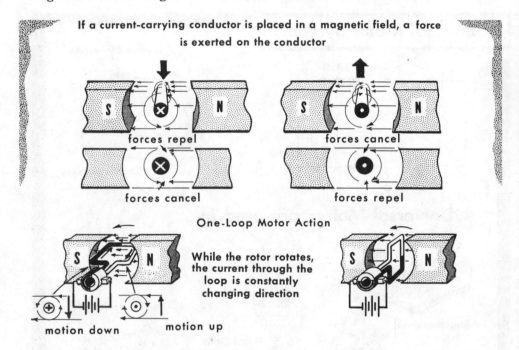

If a current-carrying conductor is placed in a magnetic field, a force is exerted on the conductor

forces repel   forces cancel

forces cancel   forces repel

One-Loop Motor Action

While the rotor rotates, the current through the loop is constantly changing direction

motion down   motion up

D-c operation differs from a-c operation in having no alternating fields. How then are we to set up a field which causes the rotor to turn? You may recall that if a current-carrying conductor is placed within a magnetic field, a force is exerted on the conductor, tending to push the conductor out of the field. On this principle, the d-c operation of the universal motor can be explained. The magnetic field is set up by the d-c flowing through the stator winding. Now, if we send d-c through the rotor, another magnetic field is set up around the rotor. As a result, a force is applied to the rotor, which then rotates.

If we consider the rotor winding to be a single loop of wire, the d-c current flows in one side of the loop and out the other. Thus, although the source is d-c, we have, in effect, a-c flowing through the loop. So, as our rotor rotates, the current is constantly changing direction within the magnetic field set up by the current through the stator winding. This creates the effect of having a reversing field within the stator, as in a-c operation, and our motor continues to run.

Commutation

A universal motor using only a single-loop rotor is not very efficient. In fact, such a rotor turns so slowly that, even though the current continually reverses itself, the rotor will probably stop after the first revolution. This will occur at the time when the north pole of the rotor and the south pole of the stator are adjacent to each other. No torque is applied to the rotor, hence no rotation results. This obvious effect is overcome through the use of a special type of rotor, called an armature. The armature, just like the stator, consists of steel discs. Each disc is shaped as shown in the head-on view below. The small holes, numbered 1 through 12, near the perimeter of the disc are called slots. When several of the discs are pressed together and properly aligned, these slots form long grooves into which are placed the armature windings. Now, we have an assembly consisting of many of the single loops. As shown in the illustration, a small inner circle is cut into 12 equal parts, designated A through L. These small sections are called armature segments.

To keep the armature turning, a device called a *commutator* is employed. The commutator reverses the connections to the revolving loops (conductors) on the armature at the instant the current in the conductors is reversing. Each end of the conductor loop is connected to one segment of the commutator. Each of these segments is insulated from the two adjacent segments.

Set against the commutators, and usually held by spring compression, are sliding contacts called brushes, which are so set that the current flows from one end of the conductor into the brush and out into the other end of the conductor loop. This occurs regardless of the direction of current flow in the conductor. Hence, current flows in one direction only through the brushes.

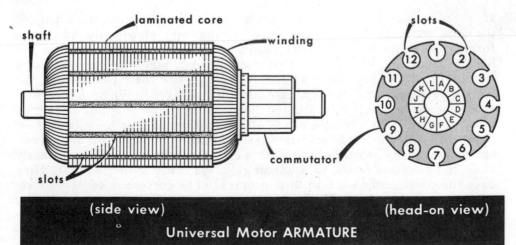

(side view)   (head-on view)

Universal Motor ARMATURE

Commutator Action in the Universal Motor

The figure below serves to illustrate the commutator action in an a-c motor. Let the outside circuit at A have a voltage which causes a current to flow in the armature loop as indicated by the arrows. Remember that any coil or loop which carries a current is surrounded by a magnetic field produced by that current. This now occurs in the armature, with the armature field having north and south poles which are perpendicular to the plane of the armature. However, the S pole of the field winding attracts the north pole of the armature, while the N field-winding pole attracts the south pole of the armature. Since the armature is free to rotate, it revolves.

COMMUTATOR ACTION in the UNIVERSAL MOTOR

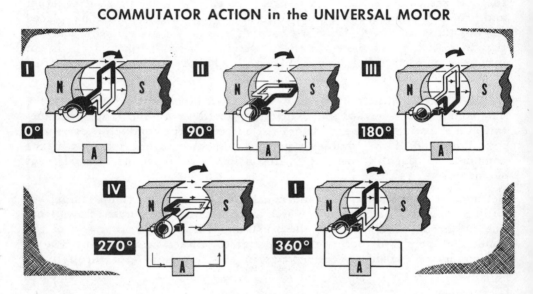

If, at the instant the north pole of the armature is adjacent to the S pole of the field winding, the armature were moving very slowly, the attractive force between the N and S poles would be sufficient to hold the armature fixed at that point. However, at that instant, the commutator segments have rotated to where they are now electrically opposite their positions at the start of rotation. Thus, the armature current is reversed. We have the south pole adjacent to the S pole, and repulsion occurs. The armature continues to revolve, with the cycle repeating itself continuously as long as current flows in the armature and field windings. This also holds true for a-c universal motors, with one exception. With a-c applied, the current is constantly changing direction in the field windings. This leads to the thought: How does the commutator work in this instance? The current is constantly reversing anyway, so why use the commutator? The answer is very simple: Both the armature and field windings have their current reversed simultaneously. The net effect of this is as though direct current were still applied to the commutator, and the above discussion is still applicable.

Speeds of Universal Motors

Universal motors are series-wound. Their no-load speed is very high. No-load speed is the speed reached by the rotor or armature when it rotates freely within its bearings with no restraining force (load) attached. Universal motors are variable-speed motors. If we place an increasing load on the motor, the speed decreases and continues to decrease as the load is increased. For our motor to be a universal type, it should operate at the same speed, regardless of whether the input current is a-c or d-c, providing the load and voltage across the motor remain constant. Since this is very difficult to obtain on very-low-horsepower low-speed motors, most universal motors are designed to operate at high speeds. These high speeds are 3500 rpm and higher. For this reason, any so-called universal motor designed for low hp at low speeds is not truly a universal motor. It is true that it operates on both a-c and d-c, but its variation in speed with a given load effectively cancels the meaning of the word "universal." Any motor is

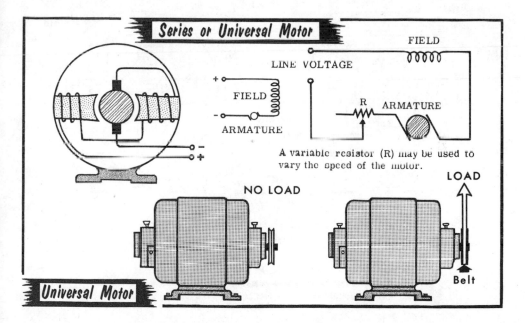

termed universal if it operates satisfactorily when the power source is varied over the rated frequency range at rated voltage and when the rated d-c voltage is applied.

The speed of a universal motor can be made adjustable. A variable resistance placed in series with one lead of the motor provides sufficient control of the input current. This, in turn, controls the amount of magnetic flux within the motor and the speed of rotation. For example, a controllable universal motor is used in a sewing machine.

Types of Universal Motors

There are two general types of universal motors — the compensated type (distributed field) and the noncompensated type (concentrated pole). By the word *compensated*, we mean that the motor has an extra winding embedded in the poles of the stator, which is connected in series with the armature. The magnetic field set up by this "compensating winding" neutralizes the cross-magnetizing field set up by the armature. This neutralization thus aids in maintaining more nearly constant motor speed. Therefore, a noncompensated motor is one without the compensating winding and with poorer speed regulation than a motor with compensation.

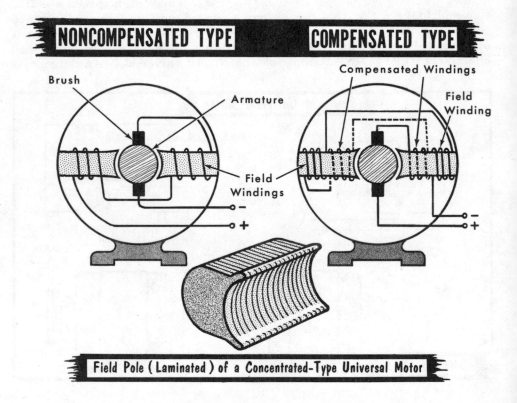

Field Pole (Laminated) of a Concentrated-Type Universal Motor

The concentrated-pole type of universal motor resembles a d-c series-wound type of motor, with one exception. In this type of universal motor, the stator is made up of laminated pieces, rather than solid pole pieces. The laminations are necessary because the magnetic field in the stator is constantly alternating when the motor is operating on a-c. If it were not laminated, excessive losses of flux would occur in the poles as a result of eddy-current effects. Concentrated-pole construction is used for the low-horsepower universal motors which operate at high speeds.

The Concentrated-Pole Universal Motor

In the figure below, the parts which make up a typical concentrated-pole universal motor are illustrated. This type of construction resembles that of a d-c series-wound type of motor, except that the stator core 1 is made up of laminated punchings, with the two field poles 2 and 3 as part of these punchings. Coils 4 and 5 are form-wound. They are placed over field poles 2 and 3 and are held in place either by iron straps, pins, or cotton tape. There are two wires coming from each of the coils 4 and 5.

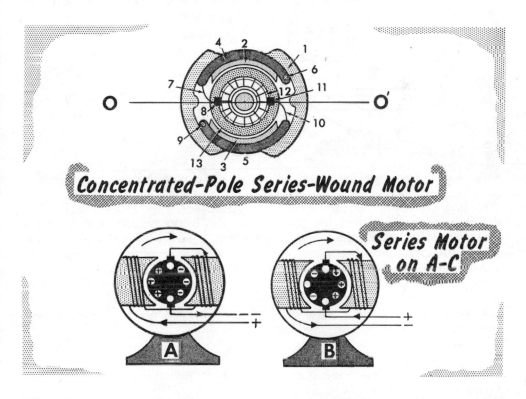

Concentrated-Pole Series-Wound Motor

Series Motor on A-C

Wire 6 connects to one side of the input power line, while wire 7 connects to the carbon brush 8. Wire 9 connects to the other side of the input power line, while wire 10 connects to the other carbon brush 11. The carbon brushes 8 and 11 make contact with the segmented commutator 12. Number 13 represents the outer diameter of the armature. Line OO′ represents a line mechanically 90° from the centers of the field poles 2 and 3. As can be seen from the figure, the carbon brushes 8 and 11 are placed on this line.

When the motor operates on a-c, both the armature and field-coil currents are reversed at the same time. This presents the same conditions as if the motor were operating on d-c and someone were continually switching the polarity of the input power lines.

The Concentrated-Pole Universal Motor (contd.)

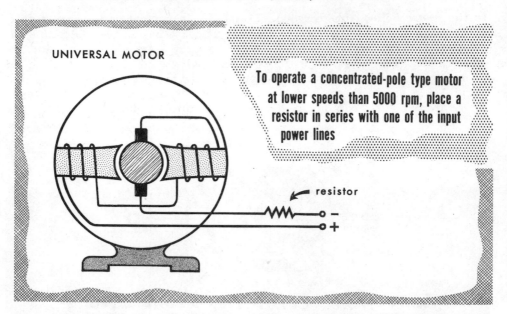

UNIVERSAL MOTOR

To operate a concentrated-pole type motor at lower speeds than 5000 rpm, place a resistor in series with one of the input power lines

resistor

Unless this type of motor is run at high speeds, say, above 5000 rpm, a different type of construction than that shown in the preceding illustration must be made. Resistance must be placed in series with one of the input power lines, as shown below. Without this resistance in the line, the motor would not operate properly on d-c or on low-frequency a-c, because a "reactance voltage" exists when the motor operates on a-c. This reactance voltage varies with the frequency of the input a-c power and has a higher value at higher frequencies and a lower one at lower frequencies. This can be seen by the use of the following mathematical formula.

$$X_L = 2\pi FL$$

where X_L is inductive reactance (ohms), L is inductance (henries), F is frequency (cps), and π is equal to 3.14.

If, for example, the inductance of the field winding is 10 henries, the following equations show how the higher frequency causes a higher reactance.

(At 25 cps:) $X_L = 2\pi FL = 2(3.14)\ (25\ \text{cps})\ (10) = 1570$ ohms

(At 60 cps:) $X_L = 2\pi FL = 2(3.14)\ (60\ \text{cps})\ (10) = 3768$ ohms

This gives a higher reactance voltage, which, in turn, affects the speed of the motor. The greater the reactance voltage, the lower is the speed, and vice versa. So, since a d-c input does not result in any reactance voltage, the extra resistance must be added to the line to keep the motor speed to the same rate as if it were running on an a-c input.

UNIVERSAL MOTORS

The Concentrated-Pole Universal Motor at Low Speeds

Some of the small series-wound universal motors are constructed with a resistance connected across the carbon brushes. In effect, this resistance forms a parallel connection across the armature — a shunt. Any parallel connection causes the current in the circuit to increase and divide up. Therefore, although the current which normally flows through the armature is still the same, the additional current now flows through the parallel resistance. The sum of these two currents flows through the field coils. This changes the ratio between the current flowing through the field coils and the current flowing through the armature, with more current flowing through the field coils. A typical numerical example is illustrated here.

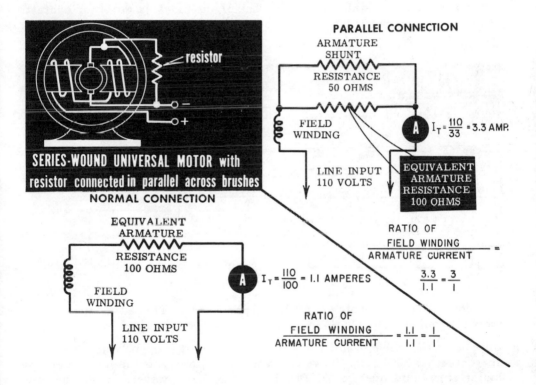

SERIES-WOUND UNIVERSAL MOTOR with resistor connected in parallel across brushes

The net effect is to decrease the motor's no-load speed. The lower the parallel resistance, the lower is the no-load speed. This controlling resistance is placed in the motor circuit to prevent the no-load speed from becoming excessive when the motor is used only intermittently. For example, many office devices, such as electric typewriters and adding machines, use this type of construction. If the resistance were not added, the motor would probably burn out during the periods when the machines are not in use but still turned on.

Control of the No-Load Speed of the Universal Motor

In addition to connecting a resistance in parallel, other means are used to hold the no-load speed of the motor. The noncompensated universal motor also uses a governor device which controls the amount of current flow. The basic idea consists of opening and closing the circuit to the stator while the motor is in operation. Centrifugal force is used to open or close the governor contacts. This centrifugal force is obtained from the motor rotation, so that, in effect, the motor controls itself. The governor unit is arranged and adjusted on the motor so that, at any preset speed, the governor contacts break, opening the stator circuit. Below this preset speed, the governor contacts remain closed.

In the control circuit shown, adjustment of the movable governor contacts is usually of the setscrew type. The setscrew is used to adjust tension of

A GOVERNOR

is used to control the no-load speed of the universal motor

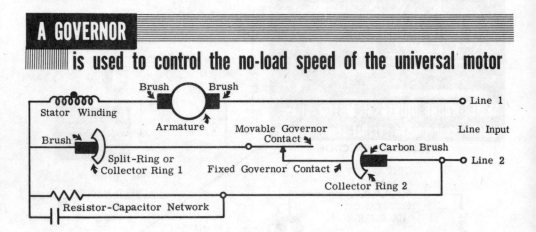

the movable contact, which is made of spring steel. This adjustment of the movable contact controls the no-load speed of the motor.

Suppose we set the movable contact tension so that it opens when the motor speed reaches 3000 rpm. If we connect lines 1 and 2 to the line input, the motor starts and operates at full line voltage. However, the instant the motor speed reaches 3000 rpm, the centrifugal force of the spinning armature forces the movable contact in an outward direction, opening the circuit. With our circuit open, the resistor is placed in series with the stator winding and line 2. The voltage drop across the resistor causes the voltage across the stator to drop, and the motor slows down. This, in turn, causes the governor contacts to close, and the cycle starts all over again. Hence, the no-load speed can never go beyond 3000 rpm if the governor is functioning properly. The capacitor ensures that the governor contacts do **not** become pitted or burned due to arcing.

Control of the No-Load Speed of the Universal Motor (contd.)

Another method, called the brush-shift method, is also used to maintain a preset load speed. It can also be used to obtain similar motor speed characteristics, whether the input is a-c or d-c for any particular load. Brush-shift is the process of actually moving the position of the motor brushes away from the neutral position. There are two neutral motor positions — mechanical and electrical. The mechanical neutral lies along a diameter line drawn through the center of the armature, perpendicular to the armature axis. The electrical neutral is determined by the connection of the arma-

The BRUSH- SHIFT METHOD is used to maintain a Preset Load Speed

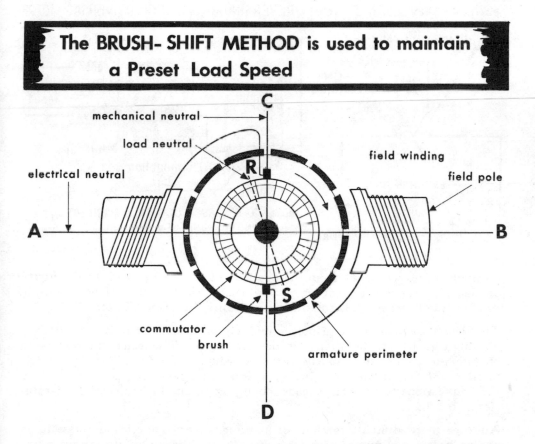

ture coils to the commutator. In the illustration, the electrical neutral has been drawn through the center of the pole pieces 90 mechanical degrees away from the mechanical neutral.

There are two electrical neutrals: the stalled neutral where the motor torque is zero; and the load neutral where the motor is operating at its rated speed, under full rated load, with minimum or zero arcing between brushes and commutator.

Control of No-Load Speed of the Universal Motor (contd.)

To utilize the brush-shift method of controlling no-load speed, we must shift the position of the brushes around the commutator in the direction of rotation of the commutator. However, care must be used in this procedure. As explained before, with the brushes set at load neutral, minimum arcing occurs. As we move away from load neutral, arcing increases and may become excessive; and we may reach a point where the armature stops turning completely. Zero speed is reached when the brushes are moved 90

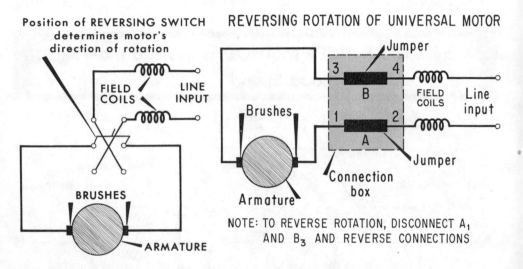

Position of REVERSING SWITCH determines motor's direction of rotation

FIELD COILS

LINE INPUT

BRUSHES

ARMATURE

REVERSING ROTATION OF UNIVERSAL MOTOR

Jumper

3   4

B

FIELD COILS

Line input

Brushes

1   2

A

Jumper

Connection box

Armature

NOTE: TO REVERSE ROTATION, DISCONNECT A₁ AND B₃ AND REVERSE CONNECTIONS

mechanical degrees away from the position of stalled neutral. In the figure on page 65, this is the position of the brushes on line RS. If we continue to shift the brushes, the armature reverses its direction of rotation.

This last statement is the principle underlying reversible universal motors, although actual brush-shifting is not employed. The usual method is to change the connections to the brushes, which, in effect, is shifting the brushes 180° from their original position. The reversing switch changes the connections to the brush leads, changing the motor's direction of rotation.

Another more difficult method of making reverse operation possible is shown here. It is much safer motorwise, since the motor direction is not subject to change as often as with the reversing switch because, once the connections are made, a partial disassembly of the motor is necessary to make reversal connections. With the extra labor involved, we are less apt to keep changing direction of rotation, causing less wear and tear and cutting down the possibility of the motor's turning one way when we expect the reverse to happen. Reversing motors are usually used in such applications as motor-driven screwdrivers, nut-setters, and intake-exhaust fans.

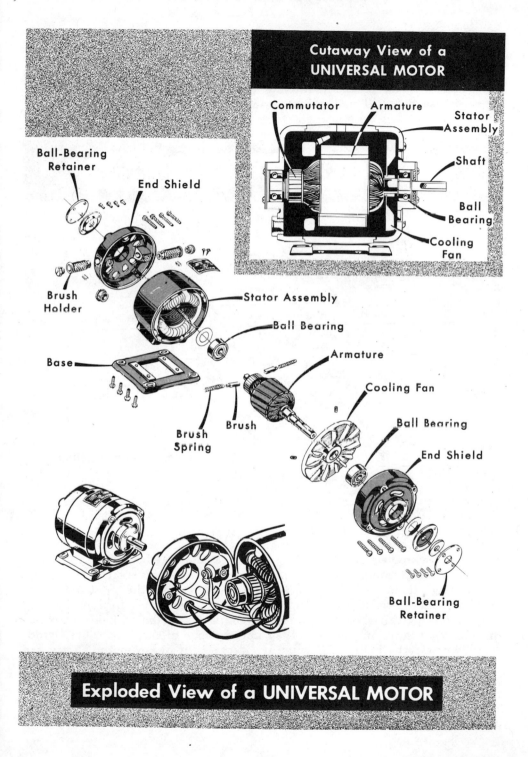

Cutaway View of a
UNIVERSAL MOTOR

Commutator   Armature
                          Stator
                          Assembly

                          Shaft

                          Ball
                          Bearing

                          Cooling
                          Fan

Ball-Bearing
Retainer

End Shield

Brush
Holder

Base

Brush
Spring

Brush

Stator Assembly

Ball Bearing

Armature

Cooling Fan

Ball Bearing

End Shield

Ball-Bearing
Retainer

Exploded View of a UNIVERSAL MOTOR

Distributed-Field Compensated Motors

The distributed-field, compensated universal motor also has its stator made up of laminations similar in general shape to those used in ordinary induction motors. However, the field winding has a similar appearance to that of a split-phase motor. The distributed-field compensated motor is used either

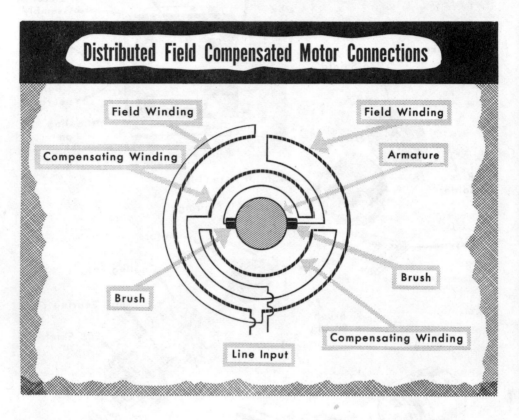

for larger horsepower ratings at high speeds, such as 1 hp at 7500 rpm, or for lower horsepower ratings at low speeds, such as 1/4 hp at 3000 rpm. The distributed-field compensated motor is more truly universal over a wide speed range than the concentrated-pole type because the previously discussed reactance voltage is compensated for. Two stator designs are employed to bring this about. In the so-called plain distributed field, the windings are distributed in slots over the complete span of the pole. These windings resemble the main field winding of the split-phase motor. They are used for the two-pole motor; however, they may be found in four- and six-pole motors. Brush-shift off the motor neutral is used to compensate for reactance voltage. The brush-shift is opposite to the direction of commutator rotation, hence such a motor cannot be used for reverse rotation; otherwise poor commutation and lower speed result.

Distributed-Field Compensated Motors (contd.)

The second type of winding used to compensate for reactance has no particular name. It consists of two separate and distinct windings. The compensating winding is distributed over the pole span. The second winding is the main field winding. This winding is placed 90 electrical degrees from the compensating winding and usually occupies two slots on each field pole. However, this type has its brushes set right on the main field-winding electrical neutral and can therefore be used as a reversible motor. Unlike the concentrated-pole type, reversal is obtained by reversing the connections to the main field winding rather than to the brushes.

In the schematic diagram of the two-winding type of compensated motor, points 1 and 9 are the inputs to compensated winding 2 and main field winding 8, respectively. Compensating winding 3 and main field winding 7 are connected, respectively, to brushes 4 and 6, which ride on the commutator 5.

If the two-winding type of motor is used for reversing service, the connections are somewhat different from those illustrated here. If you examine the connections shown in the illustration, you will notice that a reversing switch is used to switch the main field connections. Part (A) shows the direction of current when the reversing switch is in one position. Part (B) represents the reversing switch in the opposite position and the resultant direction of current flow.

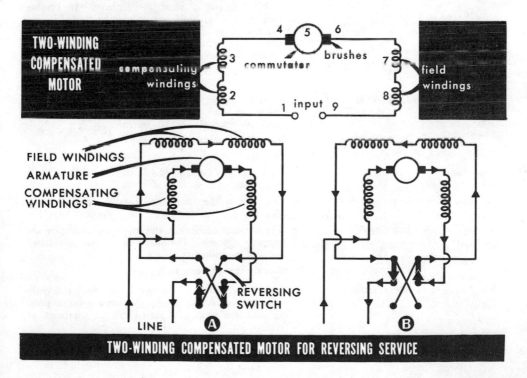

TWO-WINDING COMPENSATED MOTOR FOR REVERSING SERVICE

Troubleshooting Universal Motors

SYMPTOM	POSSIBLE CAUSE	REMEDY
Failure to start	Blown fuse or circuit breaker open	Check capacity of fuses or circuit breaker; must be rated at least 125% of full-load current.
	Low or no voltage	Check power source and see that voltage is within 10% of the voltage indicated on the nameplate of the motor.
	Open circuit	Rotate motor by hand. If motor rotates freely, check connections. Check contacts for cleanliness and tightness, and wires for breaks. An open circuit may be indicated by excessive sparking at the brushes. See that starting switch (if one is used) is closed. See that speed control governor (if one is used) is closed when motor is at rest.
	Improper line connections	Check motor connection diagram.
	Excessive load	If motor starts when disconnected from load, it may be overloaded, voltage may be incorrectly applied, or driven device may be locked or bound so it does not turn freely. Reduce load to correspond with motor rating.
	Brushes worn or sticking	Worn brushes, making only light, or no, contact with commutator, must be replaced. If brushes stick in holders, brush spring may be weak or commutator dirty. Another possibility may be high mica. Clean commutator with fine sandpaper — *never with emery!*
	Brushes incorrectly set	See that brushes are set the proper distance off neutral.
	Excessive end play	If motor fails to start, but runs if started by hand, check the shaft for excessive end play. If this occurs, add washers on shaft journal.
	Bearings frozen	(See *Overheating of bearings.*)
Excessive noise	Unbalance	Vibration can be remedied by rebalancing the rotor or pulley.
	Bent shaft	Shaft must be straightened and rotor balance checked.
	Loose parts	Check motor accessories and make sure that they are properly tightened. The motor should also be checked for firm mounting.
	Faulty alignment	Align motor properly with driven device.
	Worn bearings	(See *Overheating of bearings.*) A dry or slightly worn bearing results in slip frequency noise or purr. Oil bearings properly (periodically as required) or replace.
	Dirt in air gap	Dismantle motor and blow dirt out with dry compressed air.

Troubleshooting Universal Motors (contd.)

SYMPTOM	POSSIBLE CAUSE	REMEDY
	Uneven air gap	Check for bent shaft or worn bearings. If necessary, true up outside diameter of rotor or stator bore. Too great an increase in size of air gap may result in overheating.
Overheating of bearings	Motor needs oil	Oil motor with a good grade of oil as recommended by manufacturer.
	Dirty oil	Clean out old oil and replace with a good grade of oil.
	Oil not reaching shaft	If motor utilizes a wick, check wick and means for holding wick against shaft. If motor is of the ring type, make sure the ring rotates.
	Excessive grease	If motor bearings are ball-bearing type, excessive grease causes them to overheat. Remove excess grease.
	Excessive belt tension	Adjust belt tension to proper value.
	Rough bearing surface	Bearing must be replaced before shaft is damaged.
	Bent shaft	(See *Excessive noise.*)
	Misalignment of shaft and bearing	(See *Excessive noise — Faulty alignment.*)
	Excessive end thrust	May be caused by rotor incorrectly located on shaft or by driven device causing excessive thrust on motor bearings. Check motor and applications.
	Excessive side pull	Check application of motor to driven device. Make sure the belt tension is not excessive.
Excessive brush wear	Dirty commutator	Clean with fine sandpaper — *never with emery!*
	Improper contact with commutator	May be caused by weak brush springs or by high mica.
	Excessive load	(See *Failure to start.*)
	Governor not acting promptly	(See *Governor operating improperly.*)
	High mica	True up commutator face with a light cut.
	Rough commutator	(See *High mica.*)
Overheating of motor	Obstruction of ventilating system	Electric motors must be kept clean and dry. Inspect periodically. Remove dust; clear ventilating openings.
	Overloading	Make sure driven device does not bind. Check application by measuring input watts and amperes under normal operating conditions.
Commutator or stator burned out	Worn bearings	(See *Overheating of bearings.*)

UNIVERSAL MOTORS

Troubleshooting Universal Motors (contd.)

SYMPTOM	POSSIBLE CAUSE	REMEDY
	Moisture	Excessive moisture can weaken insulation. Replaced parts should have special varnish treatment.
	Acids and alkalis	Special varnish may be applied.
	Harmful dust accumulation	Highly conductive dust and dirt contribute to insulation breakdown.
	Overloading	(See *Overheating*.)
Governor operates improperly	Dirty commutator	(See *Excessive brush wear*.)
	Governor mechanism sticking	Work governor by hand. Replace parts if necessary.
	Worn or sticking brushes	(See *Failure to start*.)
	Low frequency in supply circuit	(See *Low or no voltage* under *Failure to start*.)
	Incorrect connections	Check connection diagram for motor.
	Incorrect brush setting	Check that brushes are set the proper distance off neutral.
	Excessive load	(See *Failure to start*.)
	Incorrect spring tension	Reset spring tension or replace springs.

UNIVERSAL MOTORS

QUESTIONS AND PROBLEMS

1. What is the principal characteristic of the universal motor?

2. Explain the d-c operation of a universal motor.

3. Why is an armature employed in the universal motor?

4. What is the purpose of the brushes in the motor? How do they function?

5. What is commutator action in a motor? With the armature rotating, explain how the commutator continues to function.

6. Why are most universal motors designed to run at high speeds? What is the true meaning of "universal" as applied to the motor?

7. Explain the difference between a compensated and a noncompensated universal motor. Which has the better operation?

8. Why is a laminated stator necessary in the noncompensated type of motor?

9. Why is a series resistance necessary for operation of the concentrated-pole type of motor when the speed is less than 5000 rpm?

10. Explain the term "reactance voltage."

11. What is the effect of placing a resistance across the brushes of the concentrated-pole type motor? What advantage is gained by this?

12. Explain why a governor device is used and its theory of operation.

13. What is the purpose of the capacitor in the governor circuit?

14. What is brush-shift? What is the difference between stalled neutral and load neutral?

15. Explain the difference between obtaining reversal on both the concentrated-pole type of motor and the distributed-field type.

16. In the distributed-field type of motor, how do we compensate for reactance voltage?

17. List some of the reasons a universal motor may not start.

18. What would occur if the governor device were to operate improperly?

Three-Phase Motors

Three major types of three-phase motors in use today are the synchronous, squirrel-cage and wound-rotor motors. The input to all three types is a three-phase voltage. Two of these motors are very similar in types of operation; the squirrel-cage and the wound-rotor types use induction as their mode of starting and running. However, the synchronous motor is usually externally started with a d-c source for rotor excitation and works on the rotating-field principle.

What makes the three-phase motor run? A stator and a rotor are required to make up a motor. Physically, the parts of a three-phase motor resemble those of other motors; that is, the stator is laminated, some of these motors have squirrel-cage rotors, the armature is just a specific type of wound-rotor, etc.

Electrically, we have a different story. The stator winding consists basically of three single-phase windings, each spaced 120 electrical degrees apart. Thus, the voltage across the three windings is 120° out of phase with both the other windings. In the typical waveform diagram shown, if we use X voltage as a reference, the Y voltage is 120° behind the X voltage, while the Z voltage is 120° behind the Y and 240° behind the X.

To connect the stator of a three-phase motor to a three-phase a-c source, two methods are used: the Y or Star connection and the Delta or Triangle connection. These are shown in the illustration. In either case, the windings are so connected that only three leads come from the stator, making the line connections very simple indeed.

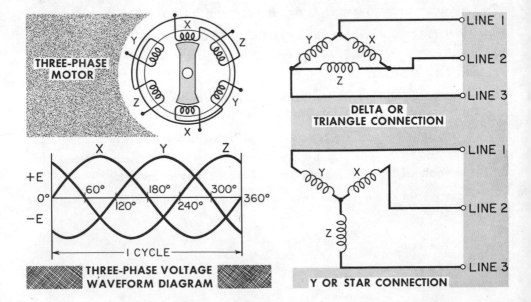

THREE-PHASE MOTOR

THREE-PHASE VOLTAGE WAVEFORM DIAGRAM

DELTA OR TRIANGLE CONNECTION

Y OR STAR CONNECTION

Three-Phase Motors (contd.)

Now, if we connect the stator to a source of three-phase a-c, an a-c voltage is impressed across the three stator windings, 120° out of phase. Referring to the illustration, let us discuss what occurs within the stator as the line voltage is applied. The directions of the current in Part (A) correspond to the point marked A in the waveform diagram. Similarly, the directions of current in Part (B) correspond to point B in the waveform diagram. Thus, in (A), we have a north (N) pole between one set of X and Z windings and a south (S) pole between the other X and Z winding.

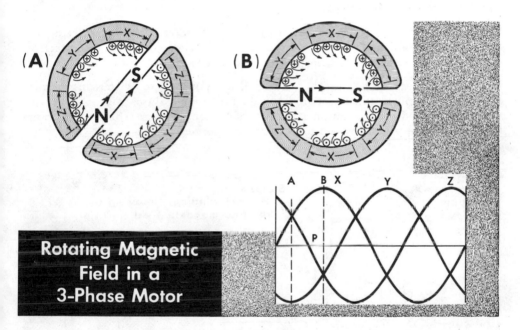

Rotating Magnetic Field in a 3-Phase Motor

Turning to Part (B), we see that the current has shifted due to an alternation in the applied line voltage. This shift takes place at point P on the waveform diagram. Close examination of the arrows in (B) shows that this current shift has also shifted the flux around the stator. Now, when the next alternation of applied line voltage occurs, the shift happens again and continues with every cycle of voltage. Thus, we have, in effect, a magnetic field moving around the inner surface of the stator. As each phase changes its current direction, the poles X, Y, and Z move with them across the full width of each phase.

Having explained stator operation, we can now turn to the specific motors mentioned earlier. We shall treat each motor in turn, rather than giving a general description of rotors, since each of the motors has a different type of rotor operation.

Synchronous Motors

The synchronous motor has the standard three-phase stator winding but uses a d-c source of excitation for its rotor. Sometimes this d-c generator, called an exciter, is mounted on the same shaft as the motor.

The applied d-c is fed into the motor with slip rings, and North (N) and South (S) poles are set up in the rotor. For example, if the motor has four poles, as shown, for every stator pole there is a rotor pole. A stator (N) pole locks with a rotor (S) pole, and vice versa. Thus as the stator magnetic field moves around the stator, the rotor moves in the same direction at the same, or synchronous, speed.

The synchronous motor does not start by itself. Some kind of starting action must be supplied. When the voltage is initially applied, the starting torque is zero, hence a starting device is needed.

One type of starter uses another motor, d-c or induction, with a high starting torque. This auxiliary motor brings the synchronous motor up to almost full speed and is automatically disconnected; and the synchronous motor comes up to full speed under its own power.

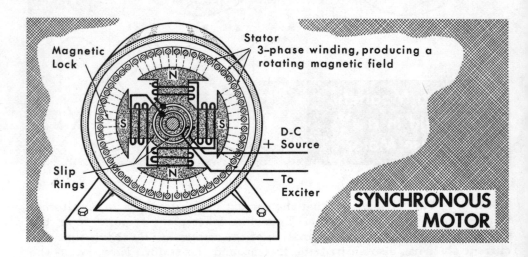

Magnetic Lock

Stator 3-phase winding, producing a rotating magnetic field

D-C + Source

Slip Rings

— To Exciter

SYNCHRONOUS MOTOR

Another method may be provided by a common winding added to the rotor d-c winding. The added winding is an induction type, or squirrel-cage construction. The voltage induced in this winding by the rotating stator field produces poles of opposite polarity in the rotor. Opposite poles attract, and we now have the necessary starting torque. At some point slightly below synchronous speed, the rotor d-c is fed in and the motor reaches full speed. Since all synchronous motors are even numbered in pole count, they are available in very high and very low speeds and can even be found as two- and four-speed types within the same housing.

Starting Small Synchronous Motors

Let us now turn to the smaller versions, which can be operated as non-excited types. We stated that the synchronous motor could not be self-started because of the tendency of the armature to pull the stator field back and forth, producing zero torque. While this is still true, the design of the small, non-excited motors is such that this condition is offset by special armature designs, or by manual start of the motor, as in the older types of electric

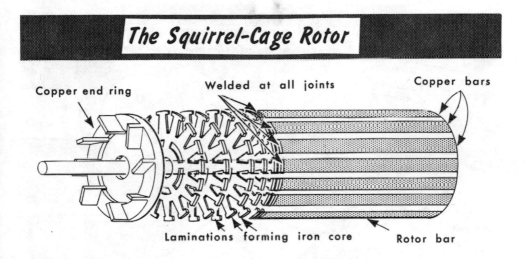

The Squirrel-Cage Rotor

Copper end ring   Welded at all joints   Copper bars

Laminations forming iron core   Rotor bar

clocks. Since only the smaller types of motors use the non-excited principle, and since the rotors are small and light, not much torque is needed for starting.

If we refer back to our illustration of the four-pole synchronous motor, we see that the poles of the rotor are all at the outer ends. Now, if instead of being made of solid bars, these poles are slotted, we may place small brass or copper bars in these slots. These bars in turn can be connected to an end ring which is similar to the end ring of a squirrel-cage rotor. With the rotor effectively short-circuited by the end ring, we have built an induction motor which starts easily with a-c, and no d-c excitation is necessary.

There are quite a few of these smaller, non-excited, synchronous motors in existence. They are used principally to operate electric clocks, recording instruments, record players, wherever a small constant-speed motor is required. On the other hand, the larger high-speed (above 500 rpm) synchronous-speed motors are used for centrifugal pumps, d-c generators, fans, blowers, centrifugal compressors. The larger low-speed (below 500 rpm) synchronous motors are used on centrifugal and screw-type pumps, electroplating generators, reciprocating compressors, and other applications where low speed and high torque are required.

Squirrel-Cage Motor

Let us turn now to the squirrel-cage motor. The principle of operation of the stator winding is the same as described in the first portion of this section. However, the type of rotor used differs from that of the synchronous motor both in construction and theory of operation.

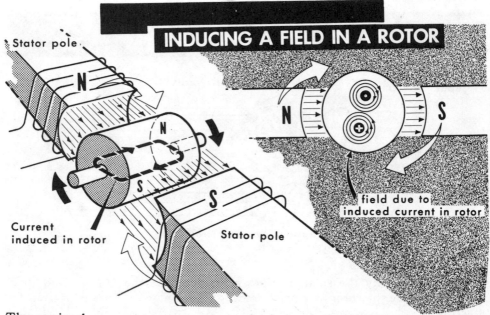

INDUCING A FIELD IN A ROTOR

Stator pole

N

Current induced in rotor

Stator pole

field due to induced current in rotor

The squirrel-cage motor operates on the induction principle. The rotating stator magnetic field induces voltages in the rotor which cause the rotor to turn. The squirrel-cage motor is so called because of its construction. The rotor consists of an iron core mounted on a concentric shaft. Copper or brass bars run the entire length of this core and are set into slots on the core. At each end of the core, end rings are welded to the copper or brass bars so that a complete short-circuit exists within the rotor. The entire assembly resembles the type of cage within which squirrels or guinea pigs are placed to run through various tests.

The squirrel-cage rotor is, in effect, the secondary winding of a transformer, while the stator acts as the primary winding. As the rotating field of the stator sweeps across the rotor windings, a voltage due to the rotating field is induced in the rotor. Since the voltage is induced, the poles produced in the rotor are always opposite in polarity to the poles of the stator field. With unlike poles, we have attraction. Therefore, with the stator poles continually rotating, the rotor poles are dragged along with them, and our motor operates.

Squirrel-Cage Motor (contd.)

Unlike the synchronous motor, the squirrel-cage motor never reaches synchronous speed. There is a simple explanation: suppose the rotor is rotating at exactly the same speed as the rotating magnetic field. It would be the same as if both the rotor and the stator were standing still with respect to each other. In that case, the lines of flux due to the stator field would merely go through the rotor, rather than be cut by the rotor bars. As we know from our basic theory, for a voltage to be induced, the lines of force must be cut at some angle other than 180°. If the rotor and the stator were standing still, the angle would be exactly 180°, hence no induced voltage. For this reason, the rotor always lags the rotating field by a small amount, thus presenting a surface which cuts the magnetic lines of force.

The speed of a squirrel-cage motor depends upon four conditions: load, applied voltage, frequency, and the number of poles within the motor.

But in the case of the squirrel-cage motor, we are more interested in the amount the rotor lags the speed of the rotating field. This difference in speeds, called slip, is entirely dependent on the load. The greater the load, the greater the amount of slip, and the slower the speed of the rotor. However, this slip is such a small fraction of the synchronous speed, that the squirrel-cage motor is used widely as a constant-speed type.

Squirrel-cage motors are not used where a high starting torque is required; however, they can be used to advantage where low and medium starting torques are needed. Because of their constant-speed characteristics, they are often used in such items as larger types of fans, conveyor-belt applications, air compressors, and presses. In the smaller sizes, they can be found in any type of motor application where a three-phase a-c source is available.

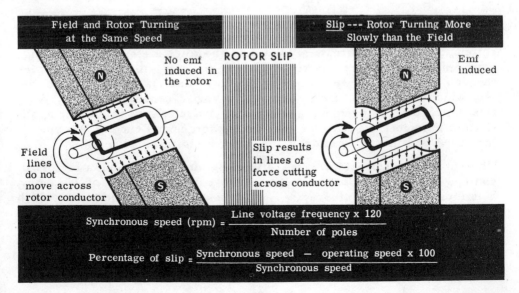

Field and Rotor Turning at the Same Speed

Slip --- Rotor Turning More Slowly than the Field

ROTOR SLIP

No emf induced in the rotor

Emf induced

Field lines do not move across rotor conductor

Slip results in lines of force cutting across conductor

$$\text{Synchronous speed (rpm)} = \frac{\text{Line voltage frequency} \times 120}{\text{Number of poles}}$$

$$\text{Percentage of slip} = \frac{\text{Synchronous speed} - \text{operating speed} \times 100}{\text{Synchronous speed}}$$

The Three-Phase Wound-Rotor Motor

Although the wound-rotor motor is very similar in principle to the squirrel-cage motor, there is one major difference in construction. Where the squirrel-cage rotor has copper or brass bars which are permanently short-circuited, the wound rotor has insulated windings in their place. These windings are not permanently short-circuited.

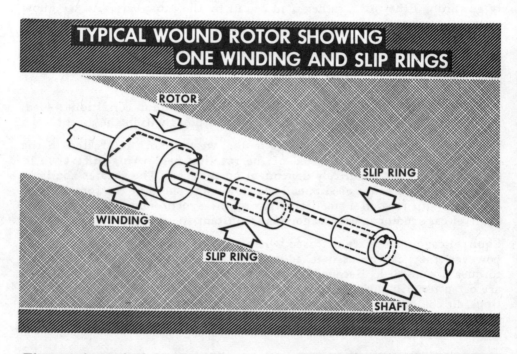

TYPICAL WOUND ROTOR SHOWING ONE WINDING AND SLIP RINGS

The starting principle of the wound-rotor motor is identical to that of the squirrel-cage motor. But the wound rotor is not permanently short-circuited; instead, the currents produced in the rotor are fed into slip rings mounted on the end of the rotor. These currents, in turn, are coupled through slip ring brushes (carbon or graphite) into an external control device which is either a variable resistor or a Variac, usually the latter. With this control device we can change the motor characteristics to suit any application. Therefore, we can vary the starting torque and current, the operating speed, and the acceleration of the motor up to full-load speed.

Though it is possible to control the wound motor's operation, a big disadvantage results. Any loss in motor speed results in a loss of efficiency. Therefore, the wound-rotor motor is used mostly on heavy equipment which requires a high starting torque and smooth acceleration up to full-rated load, or where variable speed is essential. This latter characteristic is useful in such applications as elevators, cranes, hoists, conveyors, and compressors.

THREE-PHASE MOTORS

Reversing Rotation of Three-Phase Motors

Any of the three foregoing types of three-phase motors may have their direction of rotation changed very simply. The entire procedure consists of disconnecting any two of the three stator leads from the line, interchanging them, and then reconnecting them to the line. Thus, the Z winding is made to follow the Y winding rather than the X, and the polarity is reversed. The rotating field will then rotate in the opposite direction, and the rotor will rotate along with it.

Typical connection diagrams for three-phase motors are illustrated below. Note how the stator leads are changed to achieve a reversal in rotation.

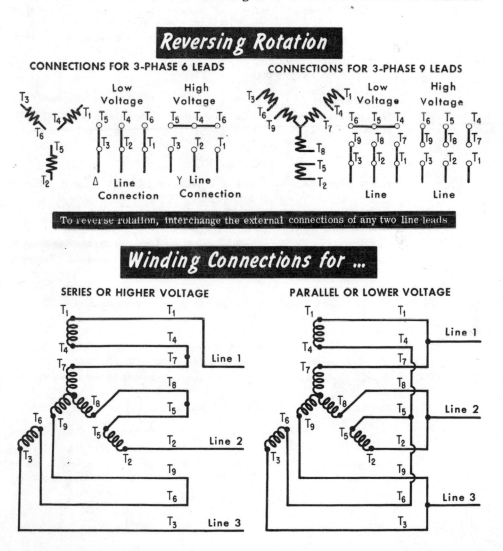

THREE-PHASE MOTORS

Troubleshooting Three-Phase Motors

SYMPTOM	POSSIBLE CAUSE	REMEDY
Failure to start	Blown fuse or open circuit breaker	Check rated fuse capacity or circuit breaker capacity. Must be rated at least 125% of full-load current. If fuse blows again, check for grounds or shorts.
	Low or no a-c field source	Check for open control devices. Check power source and see that voltage is within 10% of motor name-plate voltage.
	Bearings frozen	Check for bearings which do not rotate. Replace and lubricate as required.
	Binding load	Check load to see that it moves freely.
	Open or shorted field	Check for opens, shorts, or grounds. If these are found, field must be rewound or replaced.
	Open or shorted rotor	Check for opens, shorts, or grounds; if they occur, rotor must be rewound or replaced.

SYNCHRONOUS TYPE

	Defective exciter	To check exciter output, turn rotor by hand.
	No d-c source, exciter not used. Open exciter Variac.	Check d-c source for blown fuse or open control device. Check Variac windings for opens.

SQUIRREL-CAGE TYPE

	Dirt in air gap	Check and clean air gap. If necessary, sandpaper rotor and blow out dirt.

WOUND-ROTOR TYPE

	Broken slip rings	Check slip rings for breaks or opens. Replace if so.
	Open control device	Check windings of rheostat for opens. Replace if necessary.
Motor won't come up to speed	Low a-c source voltage	Check a-c source for proper output.
	Binding load	Free up load.
	Binding bearings	Check bearings. Clean and relubricate.
	Insufficient oil or grease	Relubricate as recommended by the manufacturer.
	Low frequency in supply circuit	Check frequency.

SYNCHRONOUS TYPE

	Low exciter output	Check exciter Variac for proper setting or shorted turns. Replace if defective.

WOUND-ROTOR TYPE

	Dirt on slip rings	Check and clean as required.

Troubleshooting Three-Phase Motors (contd.)

SYMPTOM	POSSIBLE CAUSE	REMEDY
	Broken or chipped brushes	Replace brushes or refit to slip rings.
	Improper brush contact	Check brush spring tension and readjust or replace.
	Improper Variac setting	Reset Variac as required.
Excessive noise	Unbalance	Vibration can be remedied by rebalancing the rotor.
	Bent shaft	Shaft should be straightened and motor balance checked.
	Loose parts	Check motor accessories for proper tightness and motor for firm mounting.
	Faulty alignment	Align motor properly with driven device.
	Worn bearings	(See *Overheating of bearings*.) A dry or slightly worn bearing results in slip frequency noise or purr. Lubricate bearings periodically as required, or replace.
	Dirt in air gap	Dismantle motor and blow out dirt with dry compressed air.
	Uneven air gap	Check for bent shaft or worn bearings. If necessary, true up outside diameter of rotor, or stator bore. Too great an increase in size of air gap may result in overheating.
Overheating of bearings	Motor needs oil	Oil motor with a good grade of oil as recommended by the manufacturer.
	Dirty oil	Clean out old oil and replace with a good grade of oil.
	Oil not reaching shaft	If motor utilizes a wick, check wick and means provided for holding wick against shaft. If motor is of ring type, make sure ring rotates.
	Excessive grease	If motor bearings are of ball-bearing type, excessive grease causes overheating. Remove excess grease.
	Rough bearing surface	Bearing must be replaced before shaft is damaged.
	Bent shaft	(See *Excessive noise*.)
	Misalignment of shaft and bearing	(See *Faulty alignment* under *Excessive noise*.)
	Excessive end thrust	May be caused by rotor incorrectly located on shaft or by driven device causing excessive thrust on motor bearings. Check motor and application.
	Excessive side pull	Check application of motor to driven device.
Overheating of motor	Obstruction of ventilating system	Electric motors must be kept clean and dry. This necessitates periodic inspections. Dust must be removed and ventilating openings cleared.

THREE-PHASE MOTORS

Troubleshooting Three-Phase Motors (contd.)

SYMPTOM	POSSIBLE CAUSE	REMEDY
	Overloading	Check driven device to make sure it does not bind. Check application by measuring input watts and amperes under normal operating conditions.
Rotor or stator burned out	Worn bearings	(See *Overheating of bearings.*)
	Moisture	Excessive moisture can weaken insulation. Replaced parts should have special varnish treatment.
	Acids or alkalies	Special varnish treatment may be required.
	Harmful dust accumulation	Dust and dirt which are highly conductive contribute to insulation breakdown.
	Overloading	(See *Overheating of motor.*)

QUESTIONS AND PROBLEMS

1. What are the three major types of three-phase motors in use today?

2. Name the basic similarity between the three types and the similarity in operation of two of the three types.

3. How is the stator of a three-phase motor constructed? Is there any phase difference between the windings? If so, what is the phase angle?

4. Name the two types of three-phase motor connections and the differences between the two.

5. Explain what occurs when the source voltage is applied to the stator winding. What is this effect called?

6. What is a synchronous motor? Why is the synchronous three-phase motor different from the other three-phase motors?

7. Why, basically, is the synchronous motor not a self-starting type?

8. How is the synchronous motor usually made self-starting? Explain.

9. Why is the squirrel-cage motor so called? Describe the construction of the rotor and the principle upon which it operates.

10. Why does the squirrel-cage motor never reach synchronous speed? Explain.

11. What are the four conditions upon which the speed of a squirrel-cage motor depends? Which of these conditions causes motor slip?

12. If the synchronous speed of a squirrel-cage motor is 5000 rpm and the operating speed is 4000 rpm, what is the percentage of slip?

13. What is the basic difference between the squirrel-cage and wound-rotor motors?

14. Explain the theory of operation of the wound-rotor motor as compared to the squirrel-cage motor.

15. Why is motor control seldom used on the wound-rotor motor?

16. Explain the procedure for making the three-phase motor reversible and the basic theory behind this procedure.

17. Name some of the reasons why a three-phase motor might fail to start.

18. What could cause the rotor or stator to burn out in any three-phase motor?

Enclosures and Mountings

In the present-day motor field, one can see machines of many shapes, sizes, weights, and mountings. We begin to wonder why all the diversity is necessary. Why aren't all motors which are rated for the same output and input designed so that at one look we can tell their function exactly? The answer boils down to two basic factors: motor application, and the location where the motor operates. Since there are so many types of motor enclosures and mountings, each used for a special purpose, it is impossible in a book of this size to discuss them all. Therefore we confine ourselves to a discussion of only basic type enclosures and mountings used on the motors described in previous chapters. These basic units themselves have variations which we will discuss briefly, as they apply to the motors already described.

Basically, there are five different types of enclosures in the fractional horsepower field. Their selection depends upon the job the motor has to do and the place the job is done. By presenting definitions of the types of enclosures, let us state the job and at the same time state why and where the motor is used. Our first type is the open enclosure. This type has openings in the frame, as shown. It permits both field and armature or rotor to be cooled by air flowing through these openings. It can be used in almost any application where excessive heat is no problem and where dirt and dust are kept to a minimum, for example, in such applications as office fans, outdoor compressors, sewing machines, and malted mixing machines.

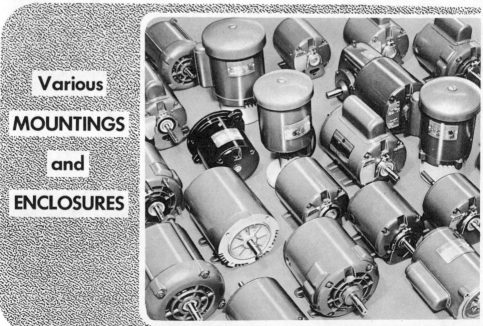

Various
MOUNTINGS
and
ENCLOSURES

The Open Drip-Proof Enclosure — Totally Enclosed Enclosure

Our second type is known as the open drip-proof enclosure. It is similar to the open type but is better protected against moisture conditions. The openings in this type of enclosure are arranged as shown below. Air can pass freely through the interior, but almost any kind of moisture dropping from

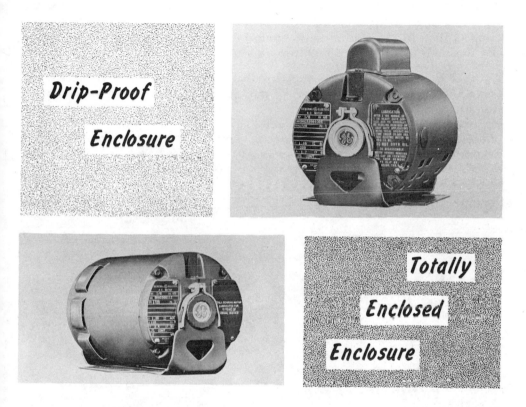

above finds it almost impossible to enter the motor. This enclosure is used in applications such as refrigeration rooms, ice-making plants, and mines, where overhead water condensation is a problem.

A third type is the totally enclosed enclosure, which has no frame openings at all, but is not sealed against water or air. There is free circulation of air within the motor, and heat is dissipated only through the frame into the surrounding cooling air. This frame must be much larger than either of the open enclosures mentioned above, since a greater surface area is required to dissipate as much heat as possible. This type of enclosure is practical only where ventilation is adequate, but where dirt or dust is a problem. For example, the totally enclosed enclosure can be found in such applications as construction equipment, machine shops, mining operations, and woodworking shops.

Fan-Cooled Totally Enclosed Enclosure — and Other Enclosures

Another type of totally enclosed motor is the fan-cooled totally enclosed enclosure. This type is essentially the same as the totally enclosed, non-cooled type except that it contains an internal fan or blower which circulates all the internal air during operation. Although running much cooler than the straight totally enclosed type, the frame size is approximately the same to allow for incorporation of the fan or blower. Its application is the same as that of a straight totally enclosed motor, but it can be used where ventilation is much more difficult to obtain.

Another type of enclosure is the explosion-proof design shown here. The frame of this type is constructed to withstand internal-gas or vapor explosion or expansion. In addition, it is so constructed that commutator sparking or arcing cannot cause explosions in external-gas or vapor-filled atmospheres. This motor is found in aircraft and ship applications where fuel vapors are a problem. Gasoline pumps, automobile starters, and the like are other similar applications.

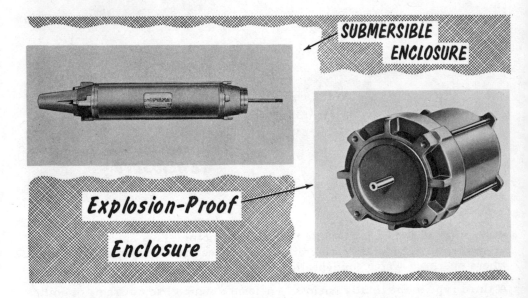

SUBMERSIBLE ENCLOSURE

Explosion-Proof

Enclosure

Some other enclosures which are essentially improvements on the basic designs are as follows: A sanitary enclosure, which is simply a totally enclosed enclosure completely air- and water-tight, for use where unwanted dust and dirt are stirred up by the motor; a weather-protected enclosure, which, as its name implies, is an enclosure used outdoors in all types of weather and is also a totally enclosed, sealed type; a submersible enclosure (pictured here) which is another totally enclosed, air- and water-tight enclosure for underwater use.

Types of Mountings

Now we come to the types of mountings upon or with which our motors and/or enclosures are fastened to their operating location. Again, there are so many different types of mountings, that we deal only with those particularly suited for use with the motors previously described.

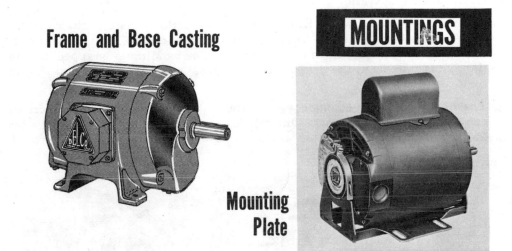

Frame and Base Casting

MOUNTINGS

Mounting Plate

By rotating end shields to keep holes upright, motors may be mounted ...

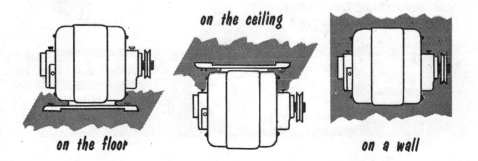

on the ceiling

on the floor

on a wall

One of the most common types of mountings is the base mount. This type of mounting is usually a plate of some sort fit to the curvature of the motor frame and mounted on the motor frame parallel to the motor shaft axis. In some motors, the base and frame are cast as one piece, while others use a type of cradle as the base mount.

The base mount can be used for ceiling, floor, or wall mounting, or in any other rigid, flat surface which can withstand motor vibrations.

Absorbant, Band, and End Mounts

Absorbant- or resilient-type mounts employ some form of resilient material, usually rubber, to cushion any motor vibration. Sometimes this type mount is simply a base-cradle mount with rubber shock absorbers incorporated into the mount's construction. Another variation is the shock mount, usually used only on small motors which may be subject to severe shock, stress, or strain during operation.

A third type, known as the band mount, is not actually part of the motor as are the base-type mounts. Any type of banding can be used, provided it can withstand motor weight and vibration. The banding is wrapped around the motor frame and the driven device or whatever surface the application requires.

The fourth, and very common, type is the end mounting. This classification can be broken down further. One type of end mounting consists of a flange on the end of the motor frame, as shown. The flange contains holes through

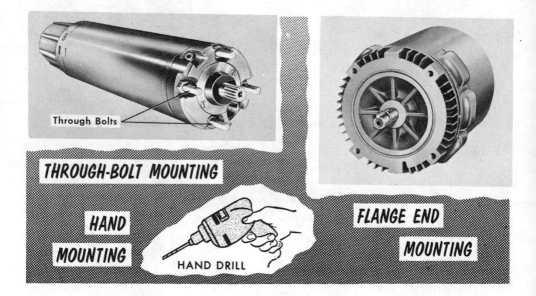

Through Bolts

THROUGH-BOLT MOUNTING

HAND MOUNTING

HAND DRILL

FLANGE END MOUNTING

which are placed mounting bolts to fasten the mount to the proper location. Some flanges have threaded studs made expressly for specific applications. These studs fit into a female flange and are secured by nuts to the threaded studs. In some cases, end mounting is accomplished through the use of longer through-bolts on the motor. These bolts go through holes in the driven device and are fastened with nuts to the applicable framework. Sometimes the motor housing itself is the mount, such as in motor-driven hand tools where the motor is encased in the housing which, in turn, also becomes the mounting.

Motor Characteristics

In order to choose a motor properly for the desired functions and to aid in determining if a motor is operating correctly, some of the characteristics to be known about a motor are: starting torque, full-load torque, power factor, efficiency, and torque-speed characteristics.

Power factor is a figure which tells what portion of the current delivered to the motor is used to do work. For instance, if we have an electric iron that draws 5 amperes, on a 115-volt a-c line, we determine the wattage by multiplying $115 \times 5 = 575$ watts. We use and pay for 575 watts. A resistive load

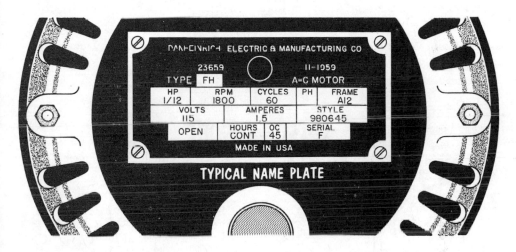

TYPICAL NAME PLATE

of this type has a power factor of 100% or unity power factor. But if we have a motor that requires 5 amperes on a 115-volt a-c line, we actually use and pay for a wattage that depends upon the power factor of the motor. A motor with a power factor of 85% utilizes $0.85 \times 115 \times 5 = 488.75$ watts. Efficiency must not be confused with power factor. Efficiency is the ratio of output power (in watts) to input power (in watts). This is expressed as a percentage figure. Thus, if the above motor is assumed capable of producing an amount of work which when converted to watts is 390, the motor efficiency is 390/488 or 80%.

Therefore, if we have a motor that has the following information on its nameplate: 115 volts; 1.50 amps; 140 watts, 1/12 hp, we can obtain the following. Total volt-amperes input = 115 volts $\times$ 1.50 amps = 172.5 watts

$$\text{Power factor} = \frac{\text{True watts}}{\text{Volt-amperes input}} = \frac{140}{172.5} = 0.811 \text{ or } 81.1\%$$

$$\text{Output} = 1/12 \text{ hp or } 1/12 \times 746 \text{ watts} = 62.1 \text{ watts}$$

$$\text{Percent efficiency} = \frac{62.1}{140} \times 100 = 44.3\%$$

Speed-Torque Curves

The speed-torque curves illustrated below compare the speed *vs* torque characteristics of the various motors shown. From these charts, we can determine just how much torque the motor will discharge at different speeds and vice versa. Thus, if the speed is known, we can determine the torque, and, from this, know whether the motor will start under the applied load. If the torque is known, we can determine the speed at which the motor will run under the applied load, and, using the chart for the split-phase motor as an example, if we know that the motor is running at 40% of synchronous speed, we simply place a ruler on the 40 line of the chart. Where the ruler crosses the torque curve, we read down and find that the percent of full-load torque is approximately 160%.

In addition, we can determine the horsepower of the machine by using our determined torque value in the following equation:

$$\text{Horsepower} = \frac{\text{Torque} \times \text{rpm}}{84,000}$$

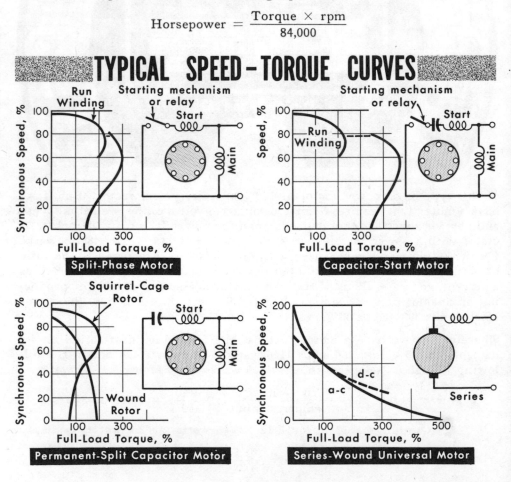

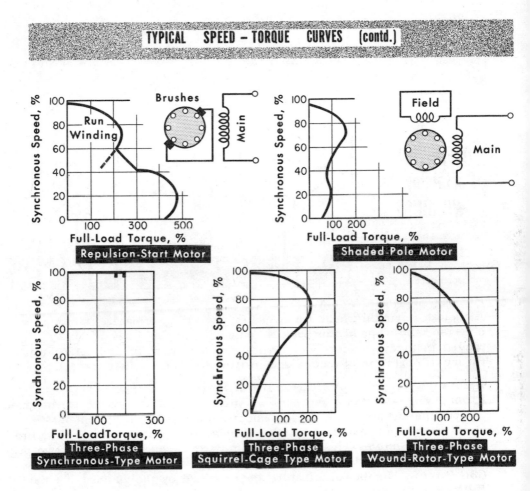

TYPICAL SPEED – TORQUE CURVES (contd.)

Repulsion-Start Motor

Shaded-Pole Motor

Three-Phase Synchronous-Type Motor

Three-Phase Squirrel-Cage Type Motor

Three-Phase Wound-Rotor-Type Motor

QUESTIONS AND PROBLEMS

1. What are the two basic considerations in choosing a motor enclosure?

2. Name the five basic types of motor enclosures in the fractional horse-power field. Describe each of them.

3. Name four other enclosures that are essentially improvements on the basic designs for specific applications.

4. Name four basic types of mountings.

5. If we have a motor with the following information on its nameplate: 230 volts; 2.60 amps; 480 watts; 1/3 hp, determine the power factor, output, and percent efficiency.

Windings

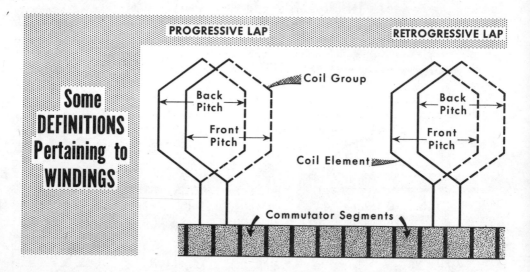

PROGRESSIVE LAP   **RETROGRESSIVE LAP**

Some DEFINITIONS Pertaining to WINDINGS

Coil Group

Back Pitch

Front Pitch

Coil Element

Back Pitch

Front Pitch

Commutator Segments

Although rewinding field, compensating, starting, armature windings, etc. is beyond the scope of this book, a knowledge of the types of windings employed in today's motors will help the reader both in understanding the theory of operation of motors and in troubleshooting these motors.

Before we proceed any further, a list of definitions follows:

Front pitch — Distance between two sides of a coil connected to the same commutator segment. Measurement is made at front of commutator.

Back pitch — Same as front pitch but measured at other end of armature.

Commutator pitch — Number of commutator segments between two segments connected to same coil.

Coil Pitch — Number of armature slots divided by the number of poles.

Full pitch — Coils which span exactly the distance between adjacent pole centers.

Fractional pitch — Any pitch which is less than or exceeds full pitch.

Coil group — Number of coils per phase per pole. Determined by dividing the number of armature slots by the product of the number of poles and phases per pole.

Coil element — One side of a coil winding.

Progressive winding — A winding where the front pitch is less than the back pitch; winding advances clockwise as viewed from commutator.

Retrogressive winding — Exact opposite of progressive winding. That is, back pitch is less than front pitch, and advance is counterclockwise.

Electrical degrees — Distance between centers of poles. (Distance between centers of two like poles is 360 electrical degrees.)

Electrical degrees per slot — Electrical area of a pole covered by one armature slot.

Types of Windings

Following is a brief description of the types of windings used in present-day electrical motors. One of the oldest types of windings employed is the Gramme-ring winding, in which a rotating ring is wound around with one single continuous coil. Taps from each wire in turn connect to the commutator segment, and the current is carried by a set of brushes. Each conductor exerts a force which tends to turn the armature clockwise. The reader can prove this for himself by examining the cross-sectional view of the Gramme-ring winding. ⊕ indicates current flow into the page, while ⊙ indicates current coming out of the page. Knowing this and from our previous discussion of one-loop motor action, we can see that the conductors carrying

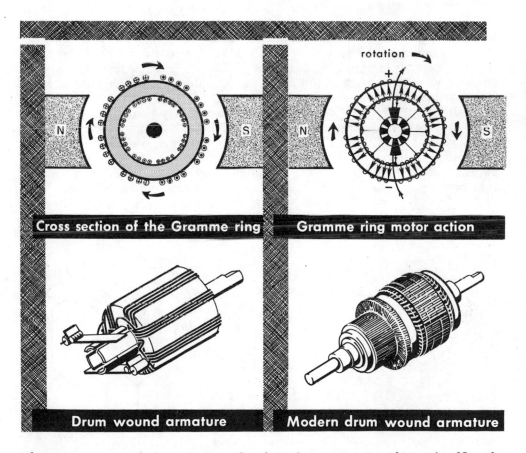

Cross section of the Gramme ring

Gramme ring motor action

Drum wound armature

Modern drum wound armature

the current out of the page are forcing the rotor away from the N pole, while the conductors carrying current out of the page are forcing the rotor down past the S pole. On practically all d-c armatures, a drum-type winding is employed. This type of winding comes in either of two classifications: lap or wave windings.

Lap Windings

Lap windings are usually used for low-voltage, high-current motors and are so called because of the way the windings lap back and forth on the armature so that one coil element lies under an N pole and one under an S pole. The lap winding is further broken down into two types: *simplex* and *multiplex*.

The lap simplex winding has only two paths from each brush, with as many brushes as poles required. A typical diagram of a four-pole lap simplex winding is shown here.

By starting at brush B and tracing the circuit successively through conductors 7, 12, 9, and 14, we reach brush C. At the same time, we see that the induced voltage in each of the conductors is in the same direction in which we trace the circuit. Now, if we start again at brush B and successively trace through the circuit formed by conductors 10, 5, 8, and 3, we reach brush A. Again, the induced voltage is in the direction in which we trace this second circuit. Knowing the direction of induced voltage and resulting current, and that current flows from negative toward positive, we see that brush B is negative and brushes A and C are positive and may be parallel-connected. Returning to the illustration, if we trace the circuit from brush C through conductors 11, 16, 13, and 18, we reach brush D. Thus, we observe that both brushes B and D are at the same potential and can also be connected in parallel. The overall total result is that even though we have four poles and four brushes, we have, in effect, only two paths through each brush.

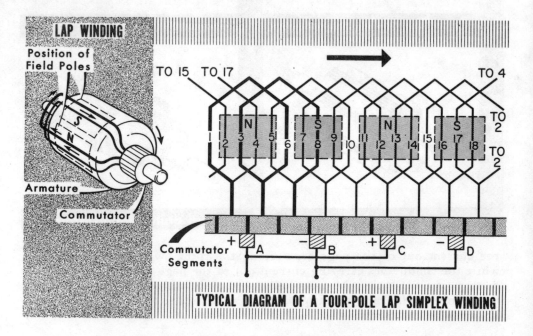

TYPICAL DIAGRAM OF A FOUR-POLE LAP SIMPLEX WINDING

Lap Multiplex Winding

Multiplex windings consist of one or more simplex windings on the armature. For example, a duplex winding consists of two simplex windings, a triplex winding of three simplex windings. These multiplex windings are used where the armature current requirements are higher than those required in the simplex type of winding. One winding of a duplex lap winding is illustrated below and is shown simply because the only difference between a duplex and a triplex winding is the connection to the commutator

Lap Multiplex Winding

NOTE: Only Odd-numbered Slot Windings Shown

segments. In the duplex winding, one winding is connected to the even numbered segments, one winding to the odd numbered segments. In the triplex winding, one winding is connected to an odd segment, the second winding connected to the first even numbered segment following the odd segment connected to the first winding, and the third winding connected to the following odd segment, and so on, depending upon the number of sections in each winding.

The multiplex winding shown has 26 segments and has 26 slots on the armature. As can be seen from the diagram, each armature slot has two layers of wire in it. If we trace any one winding from where it originates to the point where it terminates, we see that it both begins and ends in an odd numbered slot. If the other windings were shown, their ends would begin and end in even numbered commutator segments. This bears out the statement made before as to the method of connection of a duplex winding.

Lap Multiplex Winding (contd.)

Whether the type of winding be simplex or multiplex, three rules determine the number of coils used in the lap type of winding. These rules state in brief that the front and back pitch must both be odd numbers and differ by the number two or a multiple of two; the front and back pitches are laid off in different directions on the armature and are therefore of opposite sign; and the commutator pitch is equal to the average of the front and back pitches.

The winding can be one of two types: single re-entrant or double re-entrant. Both types of windings are illustrated below.

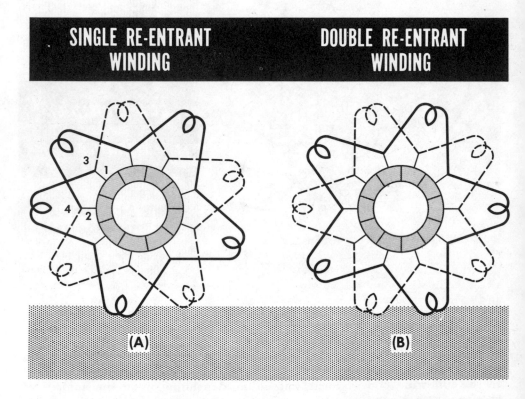

SINGLE RE-ENTRANT WINDING

DOUBLE RE-ENTRANT WINDING

(A)   (B)

As shown in diagram A of the figure, the solid winding begins at point 1 and ends at point 2. The dotted winding begins at point 3 and ends at point 4. Thus, each winding begins and ends at one point and is singly re-entrant.

Looking at diagram (B) of the figure, we see that both the solid and dotted windings do not end at any special point. This is the double re-entrant type. When the winding is double re-entrant, each brush on the motor spans at least two commutator segments.

Wave Winding

Wave windings are usually used in motors where high voltage and low current are required and are so called because of the shape of the windings on the armature. As in lap winding, the wave winding can be broken down into types: simplex and multiplex.

The simplex winding is the same as the lap simplex except that only two brushes are required regardless of the number of poles. A typical diagram of a four-pole wave simplex winding is shown in the illustration below.

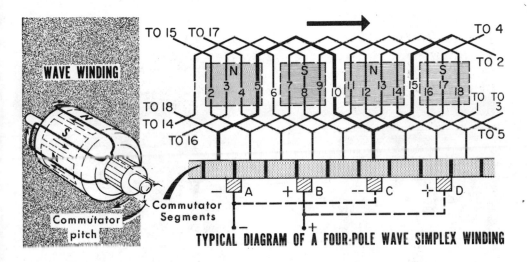

TYPICAL DIAGRAM OF A FOUR-POLE WAVE SIMPLEX WINDING

By starting at brush B and tracing the circuit successively through conductors 11, 16, 3, 8, 13, and 18, we reach brush A. At the same time, you will see that the induced voltage in each of the conductors is in the direction opposite that in which we traced the circuit. Now if we start again at brush B and successively trace through the circuit formed by conductors 6, 1, 14, 9, 4, 17, 12, and 7, we still terminate at brush A. Thus, by the same reasoning applied to the tracing performed in the lap simplex, we determine that brush B is positive and brush A is negative.

Examination of the figure shows two additional brushes, C and D, which are parellel-connected to brushes A and B, respectively. This is in contrast to what was said at the beginning of the discussion: only two brushes required regardless of the number of poles. However, further examination of the figure reveals that brushes B and D are connected only through conductors 6 and 1, with very little voltage induced in either conductor. Therefore, no other conductors are short-circuited by the connection between B and D. The same condition holds true for brushes A and C, which are connected through conductors 5 and 10 or 2 and 15. Although there are four conductors in this case, the voltage induced is negligible, and no short circuit appears.

Multiplex Wave Windings

Multiplex wave windings are constructed in a manner similar to the multiplex lap windings. That is, they consist of one or more simplex windings on the armature, such as a duplex or a triplex. However, in the case of the wave windings, the duplex winding always has four paths through the armature, the triplex has six paths, and so on. A duplex winding is illustrated here for demonstration purposes.

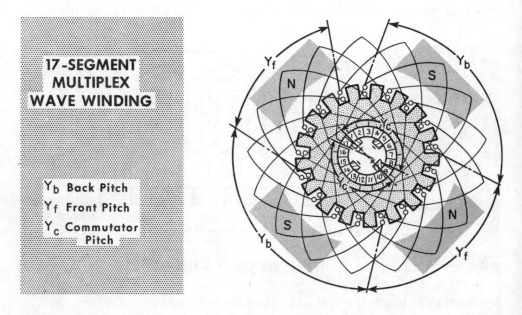

17-SEGMENT MULTIPLEX WAVE WINDING

Y_b Back Pitch
Y_f Front Pitch
Y_c Commutator Pitch

The multiplex winding shown has 17 segments and 17 armature slots. As can be seen from the diagram, each slot has two layers of wire in it. If we trace any one winding from where it originates, we see that nowhere does it definitely terminate. Thus, from the definitions of single and double re-entrant windings explained under lap windings, we have here a duplex, double re-entrant winding.

As in the lap winding, whether they be simplex or multiplex, the number of coils in the wave windings are determined by rules, four in this case. They state briefly that the front and back pitch must both be odd numbers; the front and back pitch may be equal or may differ by the number 2 or a multiple of 2; the windings are laid off in the same direction on the armature and are of like sign; and the commutator pitch is equal to the average of the front and back pitches.

On a-c motors, we also find the drum windings. In contrast to d-c motors, the drum winding is broken down into several types, of which lap and wave are simple subdivisions of major types of windings.

Distributed Windings — Chain Windings

Our first major type is the distributed winding, which has its wires for each separate phase distributed over several slots under a single pole. The three subdivisions of the distributed winding are the lap, wave, and chain types. For a-c motors, the lap winding differs slightly from the d-c lap. In shape they are the same, but they differ as to distribution in that all coils in the a-c motor form a closed circuit for every coil group. As for the wave winding, the main difference between it and the d-c wave is that only one coil in each coil group is a closed circuit.

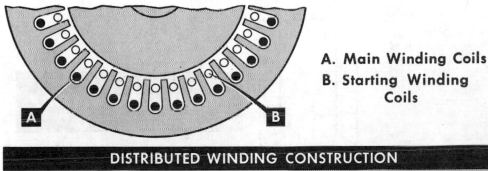

A. Main Winding Coils

B. Starting Winding Coils

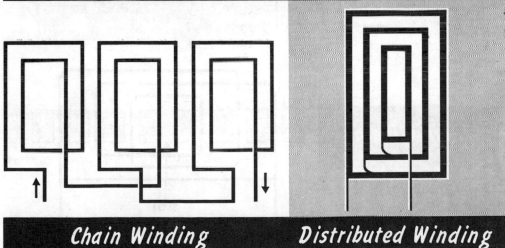

DISTRIBUTED WINDING CONSTRUCTION

Chain Winding   *Distributed Winding*

The chain winding is so called because it is wound so that it resembles the links of a chain. Since the winding also resembles a series of concentric loops, it is also referred to as a special winding. Its main characteristic is that it may be distributed unevenly; that is, it may have either an odd or even amount of wires placed in each slot. This type of winding can carry high current or withstand high voltages.

Concentrated Windings — Skein Winding

The second major type is the concentrated winding. This type is exactly opposite to the distributed type in that all wires are placed in a single slot for each separate phase under a single pole. The subdivisions under the concentrated winding heading are the lap and wave types, and they are identical to the lap and wave of the distributed winding group.

In smaller, single-phase motors, three types of field coil windings are used, skein, hand, and mold. The skein winding consists of just one skein of wires

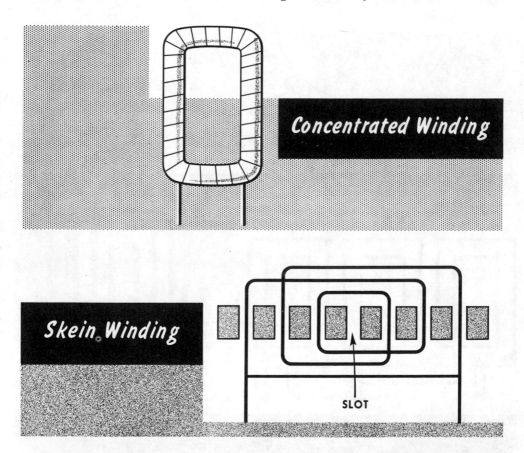

looped through the slots, forming one pole. This means that if we were to take a loop formed by several strands of wire and placed this loop in two slots, then twisted it to form a second loop, laid it in the adjacent two slots, and continued this until the first long loop could no longer be twisted, we would have a skein winding. In other words, making a skein winding is similar to taking a rubber band and continually looping it upon itself as we stretch it around some object.

The Hand or Mold Windings — Whole- and Half-Coiled Windings

The hand-type winding is exactly what its name implies. Each winding is laid in the slots by hand. It consists of a rectangular loop formed around a pole piece. As for the mold winding, it is either a hand- or machine-wound coil which is first wound on a preformed mold or casting which is usually rectangular. Then some type of insulation is baked on. The whole winding forms one solid rectangular loop. This loop is then placed in the slots and fastened into place, in the same manner as the hand-wound winding, by various types of adhesives or cements.

Single-Layer, Half-Coil Winding Having One Slot per Pole

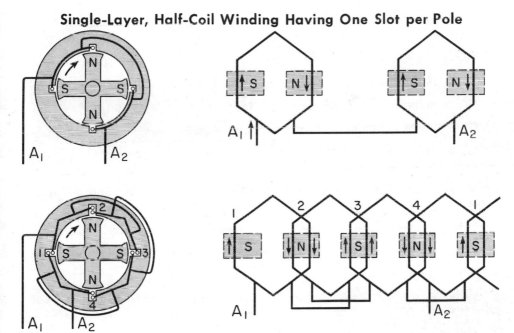

Double-Layer, Whole-Coil Winding Having One Slot per Pole

Our final major types of windings are known as whole- and half-coiled windings. By definition, a whole-coiled winding is one that is so connected that there are as many coils per phase as there are poles in the motor. A half-coiled winding is one that is connected so that there is only one coil per phase for every pair of poles in the motor. The major difference between the two types of windings lies in the method of making the coil end connections. In the whole-coiled winding, each slot contains two coil sides. In the half-coiled winding, each coil may have twice the number of turns as that of a whole-coiled winding, or the two coils under opposite poles of the whole-coiled type may be connected in series and taped or tied together to form a single coil in case of a connection change.

WINDINGS

QUESTIONS AND PROBLEMS

1. What is meant by the following terms: front pitch, back pitch, coil group, progressive winding, electrical degrees?

2. Explain the theory of operation of the Gramme-ring type of winding.

3. What type of winding is used on practically all d-c armatures?

4. Give a brief explanation of the lap winding.

5. How many brushes are required in a motor with a simplex lap winding? How many brush paths are there?

6. What is a duplex lap winding? What is a triplex lap winding?

7. What is the major reason for using multiplex lap windings?

8. What determines the number of coils used in the lap winding?

9. Give a brief explanation of the wave winding.

10. How many brushes are required in a motor with a simplex wave winding? How many brush paths are there?

11. What is the difference between a duplex wave winding and a duplex lap winding? Between the two types of triplex windings?

12. What determines the number of coils used in the wave winding?

13. What are some of the types of drum windings in present-day use?

14. What is a chain winding? A skein winding?

15. What is the difference between a whole-coil winding and a half-coil winding?

Motor Testing

Before attempting to repair a defective motor, check and inspect it thoroughly to determine exactly where the defect lies. Very often a visual check of the motor leads to the fault. Such an inspection, if conducted carefully, turns up such items as bent shafts, unlubricated bearings, worn or broken bearings, and broken leads. Quite often these require simple replacement procedures. However, if no visual checkout is conducted, futile and time-consuming procedures may have to be carried out.

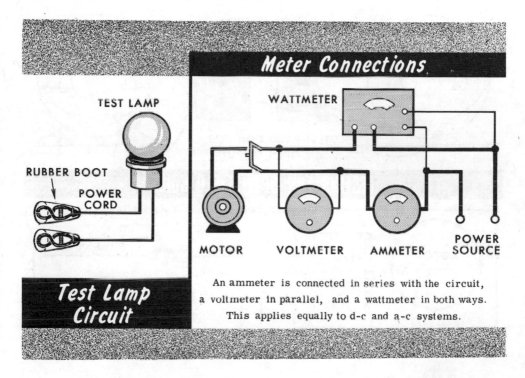

TEST LAMP

RUBBER BOOT

POWER CORD

Test Lamp Circuit

Meter Connections

WATTMETER

MOTOR VOLTMETER AMMETER POWER SOURCE

An ammeter is connected in series with the circuit, a voltmeter in parallel, and a wattmeter in both ways. This applies equally to d-c and a-c systems.

For the checking of various electrical faults which may and do occur within the motor, various tools and test equipment are available, including simple test lamps, various types of a-c and d-c meters, and less familiar items such as growlers. Even a common screwdriver may be used as electrical test equipment provided we know how to use it. Some of the more common test procedures are described in the following paragraphs.

The most common faults within a motor are "grounds" and "shorts." In a motor, a "ground" is any point on a motor component where the resistance between the component and the motor frame equals one megohm or less. A short, on the other hand, means that there is zero resistance, or extremely low resistance between two motor components, or within two parts of a component. Detection of grounds and shorts is discussed later.

Testing a Split-Phase Motor

If the motor concerned is a split-phase type using a centrifugal switch and the fault is failure to start, one or more of several conditions may exist. If the line current is connected to the motor and the shaft is turned by hand, causing the motor to start, the trouble usually lies in the starting-winding circuit. If so, the motor must be disassembled and the winding checked for an open circuit. This is done easily with a test lamp or ohmmeter.

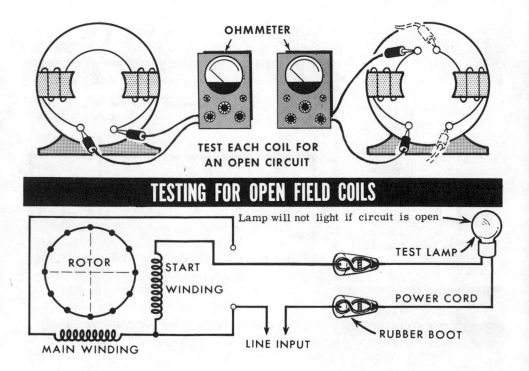

OHMMETER

TEST EACH COIL FOR
AN OPEN CIRCUIT

TESTING FOR OPEN FIELD COILS

Lamp will not light if circuit is open

TEST LAMP

ROTOR

START
WINDING

POWER CORD

RUBBER BOOT

LINE INPUT

MAIN WINDING

To use the test lamp when checking for an open circuit, connect one side of the lamp to the line input and the other side to one end of the starting winding. The other end of the line input is connected to the remaining end of the starting winding. If the lamp does not light, the winding is open and must be replaced. The ohmmeter is used in the same way you would check any coil for an open circuit. An infinite reading indicates an open circuit.

If no open circuit is indicated, check the starting switch contacts for cleanliness and freedom from pitting or burning. A small file similar to that used on automobile ignition systems can be used to correct this latter condition. If an open circuit is found, the winding may be entirely burnt out, or only one or two coils. In the first instance, the entire winding must be replaced. In the second, the coils may be rewound with new wire and then replaced.

Testing Capacitor-Start Motors

When testing capacitor-start motors, the procedures are generally the same as those used for split-phase motors. The lone exception is the capacitor. To check the capacitor's condition, several means are available. The most common one is as follows: Connect the capacitor as shown in (A) of the figure. If the fuse blows as soon as the line is connected, the capacitor is shorted. If the fuse does not blow, the capacitor may still not be any good; it may be open or operating below rated capacitance. In that case, it may be checked as follows: Connect the capacitor as shown in (B) of the figure. The meters used must be low-scale types since the current drawn from a 1-μf capacitor is of the order of 410 milliamperes. Read the indications of all three meters. Compute the power factor of the capacitor with the following formula:

$$\text{Power factor} = \frac{\text{Wattmeter reading in watts}}{\text{Voltmeter reading in volts} \times \text{ammeter reading in amperes}}$$

The power factor of a paper capacitor must not be greater than 0.1; for the electrolytic type, the power factor must not exceed 0.15. If no reading is obtained on the ammeter, the capacitor is considered open and must be replaced. If the capacitor being checked is an electrolytic type, be sure to discharge it before handling it. This may be done by disconnecting the line input and shorting the capacitor terminals with a screwdriver. A simple and quick test for a capacitor is to charge it, then discharge it, and observe the spark. On a suspected inefficient capacitor, short across the terminals with a screwdriver, and the capacitor should discharge with a *hot spark*. No spark at all indicates that the capacitor is open. A weak spark indicates a loss in capacitance.

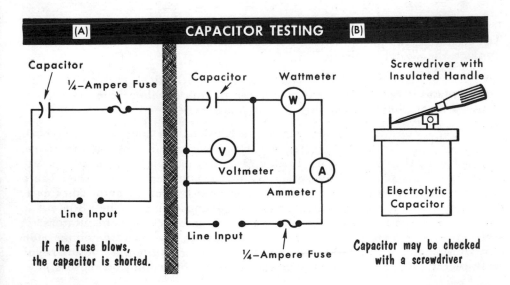

CAPACITOR TESTING [A] [B]

If the fuse blows, the capacitor is shorted.

¼–Ampere Fuse

Capacitor may be checked with a screwdriver

Testing Capacitor-Start-and-Run and Repulsion Motors

Capacitor-start-and-run motors can be checked similarly to split-phase and capacitor motors. In addition, this motor contains a two-position switch which must also be checked if the motor runs only when hand-started. When this is the case, check the end play of the shaft. If excessive, it may mean that the contacts of the two-position switch are not getting current. To

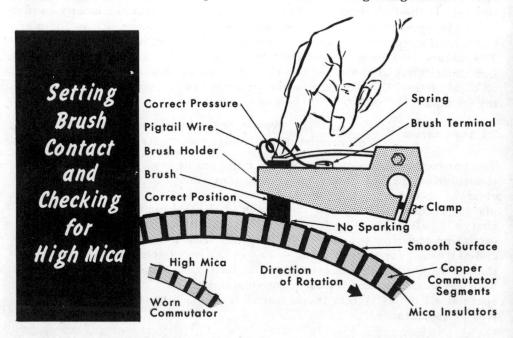

Setting
Brush
Contact
and
Checking
for
High Mica

Correct Pressure
Pigtail Wire
Brush Holder
Brush
Correct Position

Spring
Brush Terminal

Clamp
No Sparking
Smooth Surface
Copper Commutator Segments
Mica Insulators

High Mica
Direction of Rotation
Worn Commutator

overcome this, add sufficient washers on the end of the shaft to eliminate the offending end play; recheck the motor to see if it starts unaided. If not, replace the switch entirely.

In general, repulsion motors have the same mechanical faults that occur in split-phase motors. However, repulsion motors develop various electrical faults which do not appear in the split-phase type.

If the repulsion motor fails to start, the brushes, the commutator, or both, may be at fault. In the first instance, the brushes may not be making proper contact with the commutator. In this case, be sure the brushes are pushed down in their holders and the spring tension is sufficient to hold them there. In the second instance, high mica may be holding a brush or brushes away from the commutator, in which case, the commutator must be undercut as described in the section on the care and maintenance of commutators. A third reason for the motor's not starting may be a wrong line connection, which may be easily checked and remedied (refer to the diagram of repulsion-motor line connections in the section covering repulsion motors.)

Testing a Reversible Motor

If the motor being checked is an electrically reversible repulsion motor and not starting is the fault, again the cause may be the wrong line connections. Refer to the proper connection diagram to check and remedy this.

A fault occurring fairly often in repulsion motors is in the short-circuiter and brush-lifter found in repulsion-start induction-run motors. The short-circuiter should operate so that its short-circuiting action takes place with the commutator rotating at about 80% of full-load speed. If it doesn't, it may be readjusted. To ensure proper operation, adjustments must be made with the motor operating under rated load. To adjust the speed at which the weights operate, proceed as follows: Loosen screw (A). To raise operating speed, rotate threaded adjusting sleeve (B) clockwise; to lower operating speed, turn sleeve counterclockwise.

The short-circuiter may also be set at a preset speed, if so desired, but this is a difficult procedure since it entails disassembly of the motor. Nevertheless, it can be done by removing the armature together with the short-circuiting device and mounting the entire assembly in a lathe. The lathe is then operated at the desired speed while the spring (C) is adjusted with the

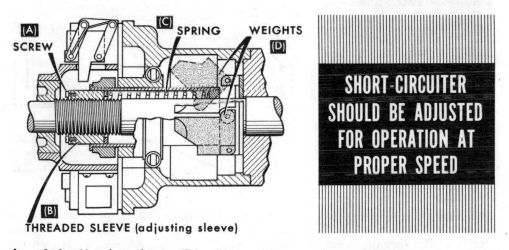

[A] SCREW   [C] SPRING   WEIGHTS [D]

[B] THREADED SLEEVE (adjusting sleeve)

SHORT-CIRCUITER SHOULD BE ADJUSTED FOR OPERATION AT PROPER SPEED

threaded adjusting sleeve (B). Since this may throw off the previous adjustment on the weights, a compromise must be made between the two adjustments until satisfaction is obtained.

Starting failures in d-c and universal motors can occur for reasons similar to those given for repulsion motors or any other motor employing commutator action. In the two-winding type of universal motor (distributed-field compensated type), the brushes may be off neutral, which, in this case, is off the compensating winding. Resetting brushes is described in the section on brush care and maintenance.

Checking the Motor for Grounds

Motor grounds exist when the insulation between winding and core permits the winding to touch metal, completing a ground circuit. These grounds can be caused by bare leads or by broken slots. Three means of determining whether grounds are present will be discussed here.

The simplest means of checking for a ground is with the test lamp mentioned earlier. The procedure consists of connecting the test lamp to the

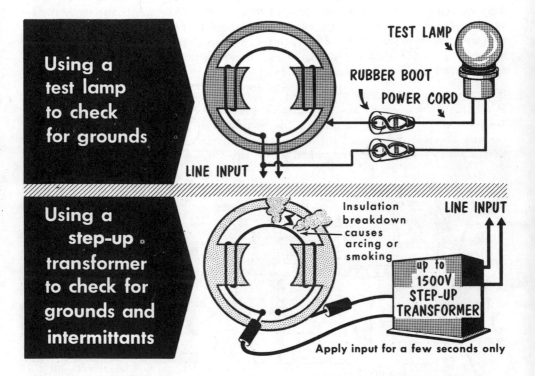

Using a test lamp to check for grounds

TEST LAMP
RUBBER BOOT
POWER CORD
LINE INPUT

Using a step-up transformer to check for grounds and intermittants

Insulation breakdown causes arcing or smoking
LINE INPUT
up to 1500V STEP-UP TRANSFORMER
Apply input for a few seconds only

winding lead and the other side of the test lamp to the motor frame. If the line input to the motor is then connected and the lamp lights, a ground exists. The line input may be either a-c or d-c but must not be kept on too long.

The test lamp method is satisfactory only where there is actual metal-to-metal contact between the winding and core or frame. If the insulation is weak, but metal does not actually touch metal, another means may be used to check for grounds. To perform this type of test, a high-voltage transformer capable of supplying up to 1500 volts is required. The test is conducted in the same manner as is the test lamp method, except that the voltage is applied by the high-voltage transformer rather than the line input. Any grounds will be indicated by arcing or smoking of the affected part.

Checking the Motor for Ground

A megger (instrument to measure very high resistance values) may be used to detect ground. The megger's leads are connected between the motor frame and the line input leads. The megger crank must be turned at a steady moderate speed. If the megger reads more than several megohms, no ground exists. If it reads less than one megohm, some portion of the coil insulation is defective, and the fault must be isolated by disconnecting the field leads and checking each coil separately with the megger.

Another type of ground may occur, for example, in the armature. It may be easily and quickly located as follows: set up the armature and testing equipment as shown here. The ground connection shown is the armature frame or shaft. Move the connection from one segment to another on the commutator. When a full-scale or appreciable deflection is obtained on the ammeter, the grounded winding is located. Correction of the ground entails rewinding the armature, which is beyond the scope of this book.

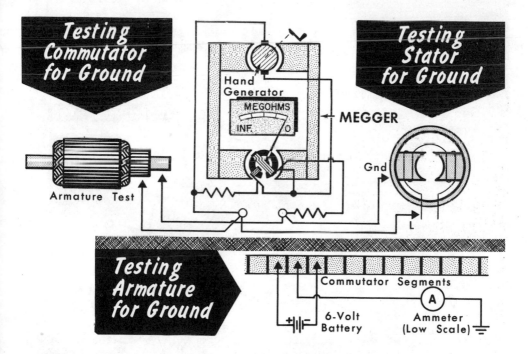

Grounds in the armature may also be located by the megger in the field-coil ground test. To conduct this test, the megger is connected between the armature shaft and the commutator segments. If the megger reads several megohms or less, the insulation in the armature has begun to deteriorate, and the armature must be repaired. If the reading is zero, a direct ground exists and necessitates immediate repairs.

Short-Circuit Tests

Short circuits may appear in both the field windings and the armature of any motor and can lead to burnout or improper operation of the motor. The methods used to detect short circuits are explained below.

An armature growler is used to check an armature for a short circuit. The term *growler* stems from the fact that, when it is used and a short circuit is detected, a vibrating noise is heard. A typical armature growler is shown here. To check an armature for short circuits, proceed as follows: (a) Insert the armature into the growler's armature holder. The adjusting nut can be loosened to permit widening of the jaws of the holder, if required. (b) Obtain a small piece of spring steel (broken hacksaw blade). (c) If the motor to be checked is a universal or small d-c, two-pole type, hold the piece of steel directly over the top slot of the armature. With the steel in place and the growler connected to the line, the vibration of the steel indicates a short circuit.

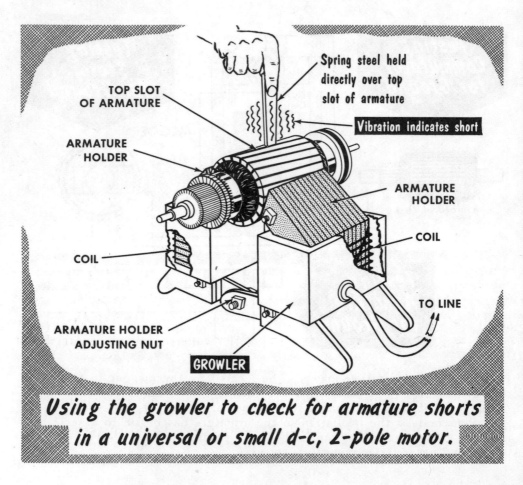

Using the growler to check for armature shorts in a universal or small d-c, 2-pole motor.

Short-Circuit Tests (contd.)

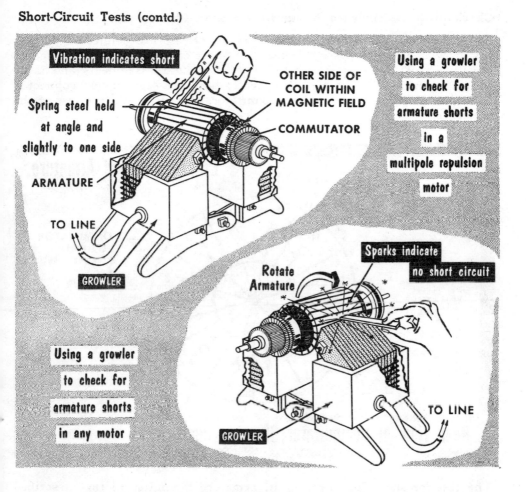

Vibration indicates short

OTHER SIDE OF
COIL WITHIN
MAGNETIC FIELD

Spring steel held
at angle and
slightly to one side

COMMUTATOR

ARMATURE

TO LINE

GROWLER

Using a growler
to check for
armature shorts
in a
multipole repulsion
motor

Rotate
Armature

Sparks indicate
no short circuit

Using a growler
to check for
armature shorts
in any motor

GROWLER

TO LINE

(d) If the motor to be checked is a multi-pole (4- or 6-pole) repulsion type, hold the steel piece at an angle and slightly to one side over the armature slot which holds the *other* side of a coil which we know to be within the magnetic field of the growler. This slot can be determined from the coil pitch of the armature. If the steel piece vibrates, the coil is shorted.

(e) If the short cannot be found by either of the above methods, check the commutator slots for traces of copper or other current-carrying particles, and clean out the slots. Another way to check for shorts with the growler is with a sharp piece of metal in addition to the growler. Turn on the growler with the armature placed in the holder. Turn the armature slowly in the holder; at the same time, touch the metal piece to the segment connected to the winding in the growler field and to adjacent segments. If sparks appear during this process, no short circuits are present. If sparks do not appear, the winding being checked is shorted.

Checking Cross-Connected Armatures for Short Circuits

Some armatures, called cross-connected armatures, are connected so that the growler cannot be used to check for short circuits. A cross-connected armature must have an even number of commutator segments and is so called because each pair of diametrically opposite segments is connected by a jumper wire in back of the commutator, between it and the rest of the armature.

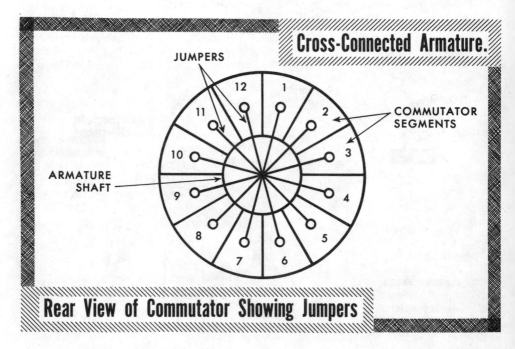

Cross-Connected Armature.

Rear View of Commutator Showing Jumpers

The test for shorts in this type of armature is similar to that described previously for armature grounds. The test apparatus is set up in the same manner except for the battery connections (see figure on page 113). For four-pole, cross-connected armatures, the battery leads should be connected to segments 90 electrical degrees apart. For six-pole types, set the battery leads 60 electrical degrees apart. Readings should be taken across all adjacent segments in this manner. If any one reading shows a deviation of more than about 8% from the average readings around the entire commutator, one of the coils attached to this segment is probably shorted. The coils must be disconnected from the segment to find the defective one.

Short circuits in the field windings of a motor can occur for any one of a number of reasons. The most common causes are a short within a coil, a short between the first and last turns of the coil which effectively shorts out the entire coil, a short circuit of an entire group of coils, or a short caused by a coil or group of coils being grounded in two places at once.

Testing Field Windings for Short Circuits

The growler is also the best means of checking field windings for short circuits. However, a different type of growler is used. Before using the growler to determine exactly in which slot the shorted winding lies, we can find the coil in which the short appears by a quick test. For this test, remove the armature and connect the field winding to the line input. Each pole in the field then becomes an electromagnet. By placing the blade of a screwdriver, in turn, on the center of each pole and then pulling it gently away, we can locate the faulty coil simply by the feel of the screwdriver as it is pulled off the pole piece. The pole with the least magnetism has the shorted coil wound around it. The same type of preliminary check can be applied to any starting windings in the motor. A smaller line voltage must be used to avoid burnout.

Once we have located the defective coil, we can use the growler to pin it down to the slot in which the coil lies. Proceed as follows: (a) Connect the growler to an a-c source (115 or 230 volts as applicable). (b) The armature having been removed in the screwdriver test, the growler can be placed within the motor housing at this point. Place growler contacts across the slots on the pole in which one coil leg is wound. (c) Place a piece of spring steel similar to that used in the armature check over the slot containing the other leg of the coil which lies in the slots covered by the growler. If the steel vibrates, the shorted coil lies in these slots. If not, repeat the procedure until the correct slots are found.

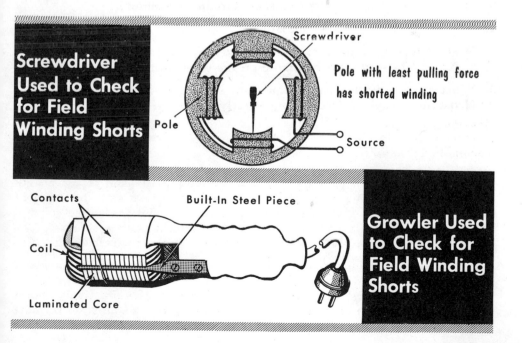

Screwdriver Used to Check for Field Winding Shorts

Screwdriver

Pole with least pulling force has shorted winding

Pole

Source

Contacts

Built-In Steel Piece

Coil

Laminated Core

Growler Used to Check for Field Winding Shorts

MOTOR TESTING

QUESTIONS AND PROBLEMS

1. What are the most common faults found within a motor?

2. What might cause failure to start in a split-phase motor? How may the fault be determined?

3. If the capacitor on a capacitor type of motor were suspected as defective, how would it be checked?

4. What would be the capacitor power factor if it were so connected in a circuit that a wattmeter read 10 watts, a voltmeter read 2.5 watts, and an ammeter read 1.5 amperes?

5. What could cause failure to start in a capacitor-start-and-run motor? How could it be remedied?

6. Explain some of the faults which may occur in a repulsion motor causing failure to start.

7. If the short circuiter in a repulsion-type motor were to fail, what would happen to the motor?

8. What is a motor ground? Where is it found?

9. What are some of the methods for determining a motor ground?

10. What are the usual results of a short circuit in a motor?

11. What is a cross-connected armature?

12. When checking a cross-connected armature for shorts, how far apart should the battery leads be placed on the commutator?

13. What are the most common causes of short-circuited field windings? Name three ways of determining exactly where the fault lies.

Care and Maintenance of Motors — Periodic Checks

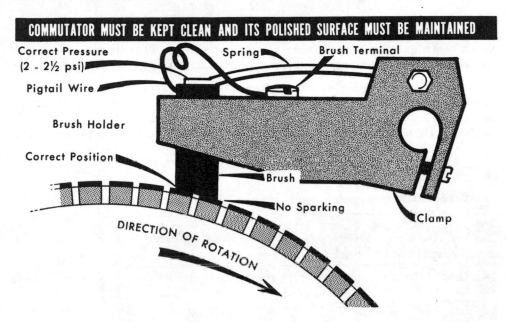

To ensure optimum operating conditions of any motor, periodic systematic checks must be performed. The frequency of these checks depends on the type of service and the operating conditions of the unit.

Keep the interior and exterior free from grease, dirt, water, and oil. If the motor is operated where dirt is at a maximum, it must be periodically disassembled and thoroughly cleaned, preferably with a vacuum cleaner. To increase the life of the motor, revarnish the windings at specified intervals, usually every one or two years, depending on operating conditions.

The rotor or commutator must be kept clean, and its polished surface must be maintained; an occasional wiping with a piece of lint-free cloth is usually sufficient. Never allow oil to get on the rotor or commutator.

On motors that utilize brushes, see that the brushes move freely in their holders and that there is good mechanical contact with the commutator. The usual pressure is between 2 and 2½ pounds per square inch. For proper maintenance, always have a spare set of brushes. When replacing brushes, be sure that they are carefully fit to the commutator.

A further periodic check for proper maintenance is to see that the insulation can stand up under high heat conditions. To do this, operate the motor for a reasonable length of time and then check the temperature of the enclosure surface. Don't use your hand for this type of check, use a thermometer. The insulation on practically all of the commonly used motors must be able to withstand up to 194°F (90°C).

Periodic Checks (contd.)

One of the most important periodic maintenance checks on the motor concerns the bearings upon which the rotor rotates. Defective, dirty, or worn bearings have caused many a motor stoppage. Several types are used in motors, each requiring a different method of care and maintenance.

The most common types of bearings employed are grease-lubricated ball or roller bearings. Usually the housings of these grease-lubricated bearings are packed with just the proper amount of grease before they leave the factory. How often the housings should be repacked depends upon the

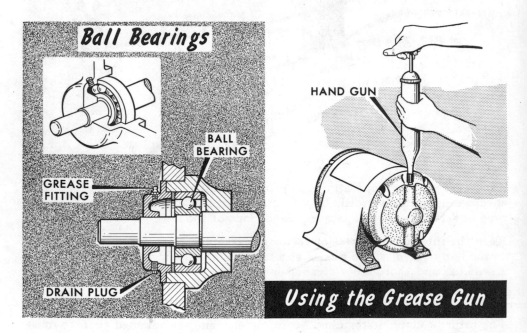

motor's operating conditions. The repacking procedure is similar to that performed in an automobile, since the grease fittings are similar in both cases. However, in the case of the electric motor, use only a hand gun to replace the necessary grease. Foot-operated or air-compressor type grease guns do not afford sufficient control over the amount of grease ejected.

Use only a high-grade grease for repacking the bearing housings. The general characteristics which such a grease should have are as follows:

(a) It should contain no abrasive matter, acid, or alkalied substances.

(b) Under normal operating conditions, or even if the motor is stored for an indefinite period, the grease must not separate into its oil and soap components.

(c) The melting point must be high, preferably above 302°F (150°C).

CARE AND MAINTENANCE OF MOTORS

Periodic Checks (contd.)

(d) The grease must have a slightly stiffer consistency than ordinary Vaseline. This consistency must be constant over the entire operating temperature range of the motor.

(e) Usually the motor manufacturer specifies the proper grease in service bulletins. Otherwise, the above conditions serve as a good guide for a proper lubricant.

To relubricate the bearing housings, proceed as follows:

(a) Clean the grease fitting and the area around it. Also, clean the area around the relief plug.

(b) Remove the relief plug and clean the relief hole thoroughly. Be sure that any old hardened grease is removed completely.

(c) Start the motor.

(d) Add grease slowly with hand gun until grease begins to emerge from relief hole. Continue to add grease until grease coming out of relief hole is clean. This means that you have pushed out most of old dirty grease.

(e) Permit the motor to run a few minutes longer until grease no longer emerges from the open relief hole; then replace the relief plug.

TABLE OF PERIODIC INSPECTION

Nature of Inspection	Daily	50	100	200
Check operation	X			
Inspect external wiring connections			X	
Check mounting studs and nuts for damage. Make certain all are tight			X	
Check brushes for wear, oil, grease, and chipping			X	
Check brushes for loose pigtails			X	
Check brushes for sluggish movement in holder and spring tension			X	
Inspect brushes for poor fit on commutator			X	
Inspect commutator for roughness and irregular wear			X	
Inspect commutator for concentricity			X	
Inspect commutator for high mica and high bars			X	
Clean slots in commutator			X	
Inspect windings for damaged insulation			X	
Check operation of control box	X			
Check for defective switches and indicator lights	X			
Inspect control box for damage and for cleanliness			X	
Check internal wiring connections			X	
Check internal wiring for insulation breaks			X	
Inspect all resistors and capacitors for evidence of damage or burning			X	
Check all receptacles. Make certain none are damaged			X	
Check all switch wires and connections for broken insulation or broken connections			X	
Check all shunts. Make sure wires and connections are unbroken			X	

Inspection Period (Running Hours) shown in columns: Daily, 50, 100, 200

Periodic Checks (contd.)

Although the procedure just described usually eliminates most old grease from the motor, there are times when the bearing housings must be completely cleaned. This type of cleaning is usually performed when the motor is disassembled for a general inspection and/or a general overhaul and reconditioning. However, if the motor has been operating in a generally horizontal position (not more than 15° from true horizontal), complete disassembly is not necessary. Prior to disassembly, mark position of end shields so that end bells realign properly.

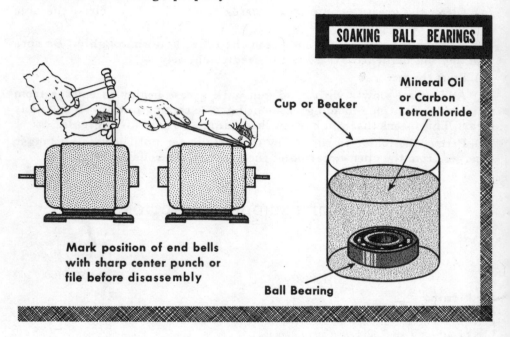

SOAKING BALL BEARINGS

Cup or Beaker

Mineral Oil or Carbon Tetrachloride

Mark position of end bells with sharp center punch or file before disassembly

Ball Bearing

Whether or not disassembly is necessary, the cleaning procedure is very similar to the repacking procedure described. However, instead of inserting a new grease, the bearing housing is flushed or soaked in either hot, light mineral oil or carbon tetrachloride. The temperature of the mineral oil should not exceed 212°F (100°C). If carbon tetrachloride is used, be sure that the area in which the cleaning takes place is well ventilated; carbon tetrachloride is toxic. Also, if carbon tetrachloride is used, reflush or resoak the housing with cool, light mineral oil to remove any carbon tetrachloride that may have been trapped in the housing. A third caution is to be sure that any carbon tetrachloride that may have accidentally splashed is immediately wiped off the winding insulation. If allowed to remain, the solution could cause insulation to rot or dissolve. Following the cleaning procedure, always repack the bearing housing with new grease as described previously.

Periodic Checks (contd.)

Another type of bearing employed is the sleeve bearing. The sleeve bearing is oil lubricated rather than grease lubricated. Relubrication of this type of bearing is very simple since it is fed through a well or plug. When this type of bearing is used, the motor is usually equipped with drain plugs.

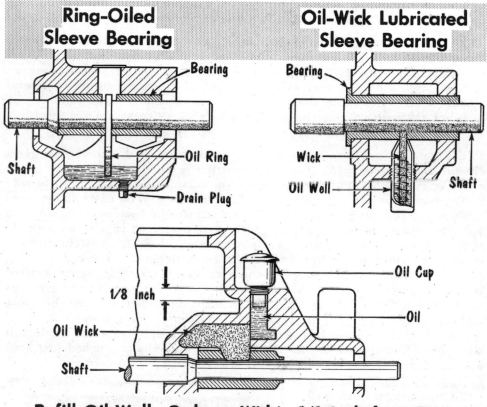

Ring-Oiled Sleeve Bearing

Oil-Wick Lubricated Sleeve Bearing

Refill Oil Wells Only to Within 1/8 Inch from Top

When so equipped, these plugs must be removed before every lubrication, so that any old oil is permitted to escape before the new oil is added. Whenever these plugs are removed, clean them in a good oil sealing compound, then replace them and tighten sufficiently to prevent oil leakage. When refilling the oil wells, refill only to within ⅛ inch from the top of the well. If the oil begins to settle after the original filling, *let it!* Adding more oil at this point won't make the motor run any better, or it may even hinder its operation. Use a good grade of oil for the relubrication. The preferred type is a light mineral oil, SAE 20.

Periodic Checks (contd.)

A third type of bearing is the *waste-packed* type. "Waste" does not refer to any type of refuse. The waste consists of a good grade of coarse wool yarn with which the bearings are packed. This type is no longer used on fractional hp sizes. Older fractional hp motors may be found with these bearings.

CAUTION: *In motors utilizing this type of bearing, the bearings are shipped dry from the factory. It is up to the motor user to pack the bearing initially before the motor is placed in operation for the first time.*

This type of bearing is also oil lubricated, as is the sleeve bearing. Motors using this type of bearing are equipped with drain plugs that must be removed before replacing any oil or when oiling for the first time. New oil is then poured over the waste packing and the bearing lining in liberal amounts. Any excess oil then flows through the open drain holes or the overflow. When replacing the drainage plugs, clean them and dip them in a good oil sealing compound to prevent any future leakage of oil.

If the motor in question is the open type, equipped with an overflow gage, the oil must be poured into the gage until it is filled. If the motor is the enclosed type without an overflow gage, the oil must be poured into the side of the bearing which is opposite the waste packing. The chart shown here enables you to pour just the proper amount into the bearing. The oil used for lubrication must be of a good grade. The preferred type is the same as that used for the sleeve bearings: light mineral oil, SAE 20. Relubrication must take place at the end of the first three-month period of operation. The procedure must be repeated every three months afterwards, or more often if conditions are severe. When relubricating, always remove the drainage plug to allow any old oil to drain off. In addition, always pick up the waste and repack it into the bearing. This procedure prevents the waste from becoming matted. At yearly intervals, remove the old waste and replace it with new. The new waste must be lubricated as if the motor had just been shipped from the factory and had not as yet been put into service.

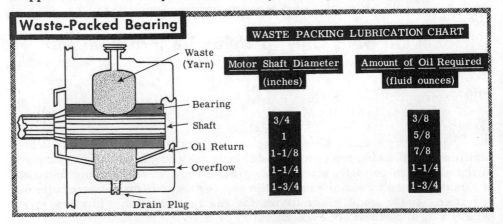

Waste-Packed Bearing

Labels: Waste (Yarn), Bearing, Shaft, Oil Return, Overflow, Drain Plug

WASTE PACKING LUBRICATION CHART

Motor Shaft Diameter (inches)	Amount of Oil Required (fluid ounces)
3/4	3/8
1	5/8
1-1/8	7/8
1-1/4	1-1/4
1-3/4	1-3/4

Commutator Care

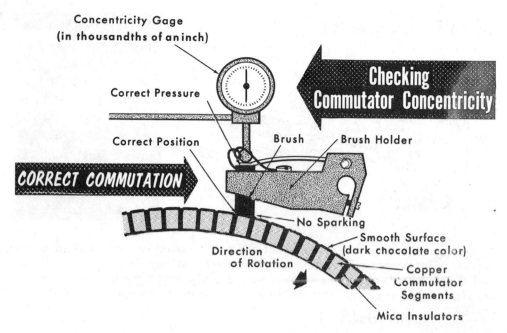

Good commutation is the most important factor in the successful operation of any commutator-type motor. By good commutation we mean operating under proper conditions without any brush burning, segment burning, or excessive sparking. For this to be maintained, the surface of the commutator must be in good operating condition — concentric and smooth, with the spaces between segments properly undercut. With overloading and overspeeding of the motor, the reverse occurs. Once a surface becomes poor, continued operation will only aggravate the condition.

By concentricity we mean that the commutator should form as nearly a perfect circle as possible when viewed from the shaft end. The concentricity should hold true for the entire length of the segments.

By smoothness we mean that the segments should not have any rough spots, hills, or depressions. By undercut we mean that the level of the insulation, usually mica, between the segments should not be at the same level as the surface of the commutator segments. Nor should the level be too low, since the deeper space would permit collection of foreign matter which may short-circuit the segment and cause motor burnout.

Returning to concentricity, for the surfaces of commutators whose speed is in the neighborhood of 5000 feet per minute, concentricity should be within 0.001 inch. At speeds of about 9000 feet per minute, concentricity should be within 0.0005 inch. This concentricity can be measured with a dial gage mounted on a brush of the type normally used in the motor.

Commutator Care (contd.)

Setting Armature up on Lathe

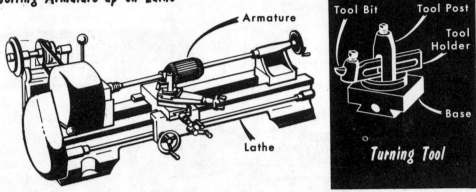

Armature

Lathe

Tool Bit   Tool Post

Tool Holder

Base

Turning Tool

It is most important that we inspect the commutator surface at regular intervals. These intervals should be so spaced that any faults found on the commutator surface can be detected before too much damage is done. If resurfacing is required, three distinct methods may be used, as follows:

(a) Turning and grinding the surface.

(b) Handstoning the surface.

(c) Sandpapering the surface.

The best method of reconditioning the commutator surface is to turn it or grind it.

If the commutator is to be turned, the armature should be removed from the motor and set up on a lathe. An additional advantage is gained if the armature shaft is first placed within sleeve bearings on each end and the bearings then mounted in the lathe.

Before turning the armature, we must provide some means of protection for the armature windings. This protection is a must since the turning operation always causes large chips of copper to fly off. If these flying chips should lodge in the windings, short circuits and burnout are almost inevitable. So, to be safe, wrap a piece of oilcloth or flexible canvas as tightly as possible around the windings, especially near the commutator end of the armature. Be sure that all possible holes are covered up.

The commutator is turned by means of a steel tool of a type similar to the typical tool shown here.

The point of the tool should be sharp, but rounded slightly to avoid leaving rough spots on the commutator during the turning process. Use small, light cuts, rather than large cuts. Greater accuracy is obtained in this manner, and the chance of tool or segment breakage is lessened.

Commutator Care (contd.)

The speed of the armature during the turning process must not exceed 600 feet per minute. However, once the surface has been turned, final smoothing must be performed at the highest speed obtainable at which the armature does not vibrate in the lathe. The final polishing is done with a stone mounted in place of the cutting tool; then very fine sandpaper is used to furnish a touchup, with the commutator turning over at maximum speed.

If the armature is to be ground rather than turned, lathe mounting the armature is not absolutely necessary. The armature can be left in the motor and run at near normal operating speed. However, the best method is still the lathe method. As in turning, the armature windings must be protected from flying particles and so must be covered in a similar manner. In grinding, the particles consist of fine copper and grinding wheel chips, so that, if possible, the protective covering must be wrapped even more securely.

The grinding procedure can be performed by either one of two means: a stationary grindstone or a rotating grindstone. The revolving grindstone is preferred because of a smaller tendency to dig into the commutator during the grinding operation. In addition, a hard-surfaced grindstone is not too satisfactory. With a hard surface, the copper dust tends to stick to it, and we eventually wind up with a grindstone as smooth as the surface we are trying to smooth out.

Again, as in turning, it is preferable to make small, light cuts rather than large cuts. Large cuts may induce faults which were not there before or accentuate already existing dents and hollows.

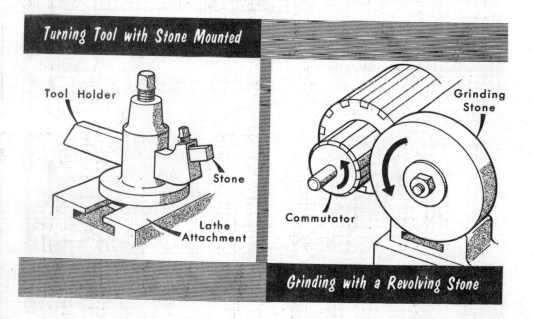

Turning Tool with Stone Mounted

Tool Holder

Stone

Lathe Attachment

Grinding Stone

Commutator

Grinding with a Revolving Stone

Commutator Care (contd.)

When using the revolving grindstone, initial grinding must be performed with the armature spinning at a rate half, or less than half, of the normal rated speed of the motor. With most of the faults eliminated, final fault elimination can be acomplished with the armature spinning at full rated speed. Final surfacing may be done in the same manner as in the turning operation already described.

The second preferred method of resurfacing a commutator is with a handstone. This type of stone usually consists of some sort of synthetic abrasive material. These stones have the property of tearing loose as soon as the particles on the stone surface become dull. Thus, the surface presented to the commutator is always sharp. Using a handstone, although inferior to grinding or turning, is superior to sandpaper. The stone presents a stiffer surface than the sandpaper and can be used to eliminate hollows, dents, and flat spots. The stone surface must be of the same curvature as the armature; that is, its concave side must match the convex surface of the commutator. During this type of operation, the armature can be either lathe-mounted or left in the motor housing. Again, lathe mounting is preferred.

When using the handstone, equal pressure must be applied to both ends and the center of the stone. For good resurfacing, the stone must be moved slowly from side to side, with the armature turning at normal speed. After the initial, coarse surfacing has been completed, the stone must be replaced with one of a finer grain to eliminate any scratches or rough edges left by the coarser stone. *Never use oil or any other lubricant with a handstone!* Oil will only clog the stone surface and decrease its grinding ability.

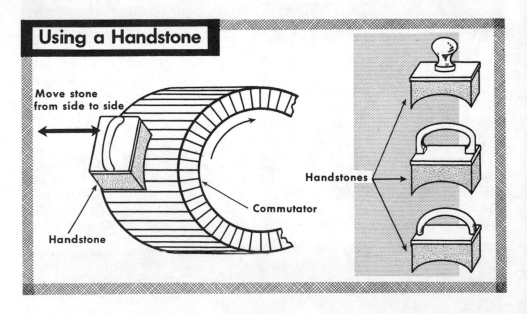

Using a Handstone

Move stone from side to side

Handstone

Commutator

Handstones

Commutator Care (contd.)

Not entirely satisfactory for use in resurfacing, sandpaper is adequate when the only faults are high mica (when the mica insulation between the segments is higher than the segments themselves), foreign matter on the commutator surface, or roughness. However, if there are hollows or flat spots in the segments, sandpaper will not remove them.

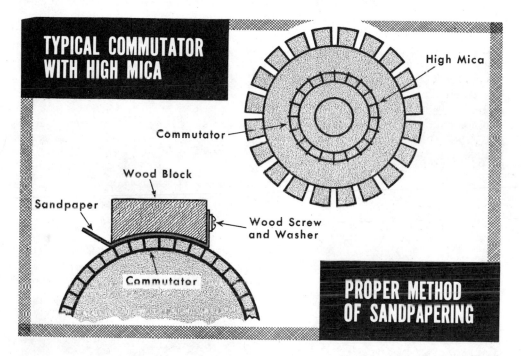

Do not use a coarse-grain sandpaper. Use a finer-grain paper and work more slowly for much better results. As a solid backing for the sandpaper, shape the surface of a block of wood so that its concave surface fits the convex commutator surface, as shown in the diagram, and fasten the sandpaper to the block. Movement of the block must be perpendicular to the axis of the segments, and very slow, to avoid scratching.

If the commutator being sandpapered is unslotted and has high mica, use a very small amount of light oil on the sandpaper to eliminate the danger of flattening the centers of the segments. If the commutator is slotted, do not use any oil. The use of oil might cause the copper and sand residue to cake together and fall into the slots, causing short circuits.

Never use emery paper or cloth to maintain a commutator! Emery particles are current carriers, and, if sufficient emery dust collects in the slots, short circuits result.

CARE AND MAINTENANCE OF MOTORS

Commutator Care (contd.)

A commutator which has been resurfaced frequently contains many fine, invisible scratches. A process called burnishing removes these scratches and leaves a highly polished, smooth surface.

Of two convenient methods of burnishing, the easiest is to use a hardwood, such as maple, held against the commutator by hand as the armature rotates. Shape the wood as described for handstone and sandpaper methods.

The second method is to use the same type of hardwood, cut into blocks the same size and shape as the brushes in the motor. These blocks are then used in place of the brushes and the armature turned by some means, possibly a belt or mechanical connection to another motor.

In the discussion of commutator care, we mentioned that the commutator must be properly undercut, that is, the mica between the segments of the commutator must be removed, so that the commutator segments wear evenly during operation of the motor. Every piece of mica must be removed to be sure that no single piece is raised above the level of the segments (high mica). If the level of the mica is even slightly higher than the segments, the brushes do not ride the commutator properly, causing arcing, spitting, and burning.

Hand Tools for Undercutting

There are many different tools used for undercutting. Choice of a tool depends on the size of the commutator to be undercut. Since we will rely chiefly on hand tools to perform the job, we will discuss only that type of tool. One very good hand tool is a slotting file (shown below). This little tool cuts a V-shaped slot between the segments and is available with either a 40° or 60° angle between the edges of the file. It can be used on motors of moderate sizes and, in addition, for beveling the edges of U-shaped or square-bottom slots. The small hook at the end of the tool is used to scrape out the mica.

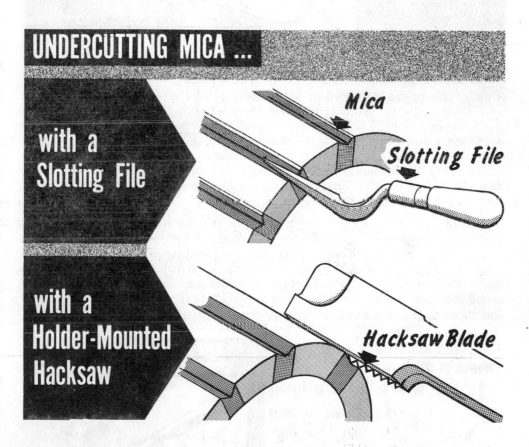

UNDERCUTTING MICA ...

with a Slotting File

Mica

Slotting File

with a Holder-Mounted Hacksaw

Hacksaw Blade

A second undercutting tool is used only when the undercutting required is comparatively simple. This tool may be hand-fashioned. It consists of a hacksaw blade fastened with a holder. This tool can be used satisfactorily only if the blade employed is always kept sharp. If the blade becomes dull, it will no longer cut the mica cleanly. Instead, it will chip the mica and may leave very thin flakes of mica along the sides of the segments.

Hand Tools for Undercutting (contd.)

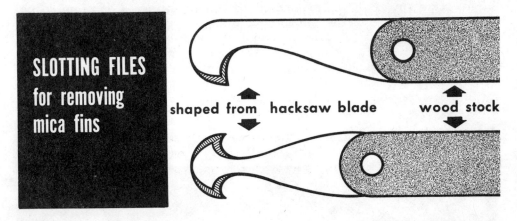

SLOTTING FILES for removing mica fins

shaped from hacksaw blade   wood stock

In addition to the U-shaped or square-bottom slot already mentioned, there is a V-shaped slot. Both types have their advantages. The U-shaped slot permits the adjacent segments to wear down to the bottom of the undercut slot before undercutting is again required. However, the V-shaped slot will not collect as much dust during operation as the U-shaped type. To retain some of this latter advantage, a U-shaped slot must be undercut not more than 3/64 of an inch. In undercutting the U-shaped slot, extreme care must be taken to avoid leaving any mica in the slot.

Any mica left at the sides of the slot is often invisible and can lead to very poor commutation. This condition is known as feathered-edge mica, mica fins, or side mica. A tool for correcting this troublesome state can be made by anyone familiar with the use of hand tools. A typical tool is illustrated here. Very easily made, it requires only an old hacksaw blade and a piece of wood stock for the handle. The edge of the blade is sharpened so that it will cut through the thin mica, as shown here.

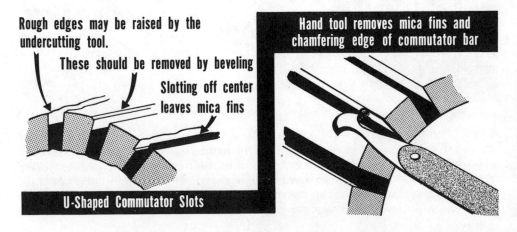

Rough edges may be raised by the undercutting tool.

These should be removed by beveling

Slotting off center leaves mica fins

U-Shaped Commutator Slots

Hand tool removes mica fins and chamfering edge of commutator bar

Hand Tools for Undercutting (contd.)

The V-shaped slots are undercut with the slotting file previously described. However, the use of this tool can also lead to some of the faults we mentioned in the discussion of U-shaped slots; that is, feathered-edge mica, mica fins, or side mica. If the slotting file is not kept in the direct center of the slot during the undercutting process, the condition as illustrated in (A) of the figure may result. If so, the same tool used to remove this type of fault for the U-shaped slot can be used with the V-shaped slot.

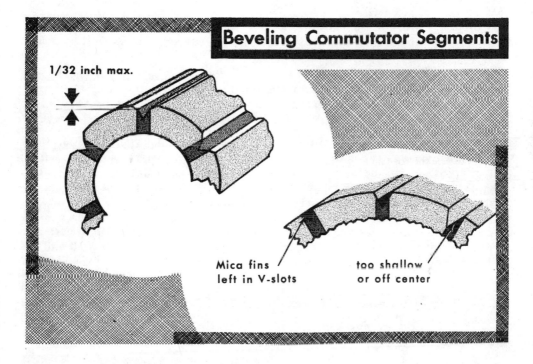

Beveling Commutator Segments

1/32 inch max.

Mica fins
left in V-slots

too shallow
or off center

Once the undercutting operation is finished, examine the edges of the segments at the slots. There is almost always a sharp edge of copper at the top of the slot. This edge can play havoc with the brushes if it is not eliminated. It is here that our slotting file comes into play again. This time, we use the file to bevel the edge of the segment and remove any chance of the edge cutting into the brushes. The bevel should be approximately 1/32 inch, and no more.

Following the beveling operation, it is a very good idea to insulate the freshly undercut slots. This can be done with a coat or two of a good insulating varnish. The varnish should preferably be of the air-drying type. Once the varnish is completely dry, any excess varnish must be removed from the segments with sandpaper or a handstone, as described previously.

Brushes

Almost as important as the condition of the commutator to the proper operation of the motor are the brushes used. Manufacturers spend a great deal of time in the selection of brushes, and for good reason. The brushes perform three separate functions, and, if they are not the proper type for the motor involved, these functions cannot be carried out properly. These functions are as follows:

(a) Brushes must carry the load current to and from the armature of the motor.

(b) Brushes must act as unlubricated bearings on the commutator, or as slip rings, as applicable, and this at surface speeds sometimes well over 800 feet per minute.

(c) On motors with commutators, brushes must safely control the short-circuit current which results from the uncompensated voltages in the armature windings being commutated. They must stand up under very high current peaks which are often a characteristic of the commutation cycle.

To perform the individual functions under different types of operating conditions, various types of brushes are used. The most frequently used is the graphite brush, because of its excellent lubricating qualities. Regardless of the type used, it is most important that they be constructed to withstand the maximum motor current without breaking down. Some of he graphite brushes, for instance, are rated at about 150 amperes per square inch of area, but these are special types of metal graphite. However, the average current capacity of most brushes lies somewhere between 40 and 60 amperes per square inch of area.

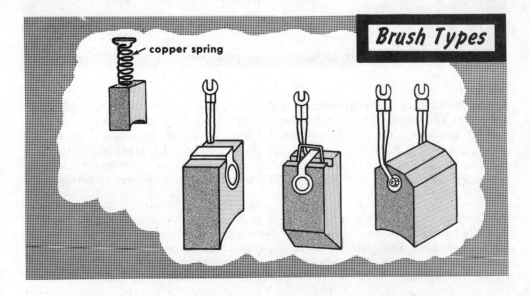

copper spring

Brush Types

Brush Holders — Reaction and Box Types

To hold the various shapes and sizes of brushes, several types of brush holders have been designed. They fall into five general categories, as listed below:

(a) Reaction type
(b) Box type
(c) Clamp type
(d) Cartridge type
(e) Rocker-ring type

Reaction-type brush holders are usually mounted so that they hold the brush at an angle of 30° or more from the radius of the commutator. Also, to avoid excessive friction and wear, they are usually placed so that the brush faces in the direction of rotation of the armature. A typical reaction-type brush holder is illustrated here.

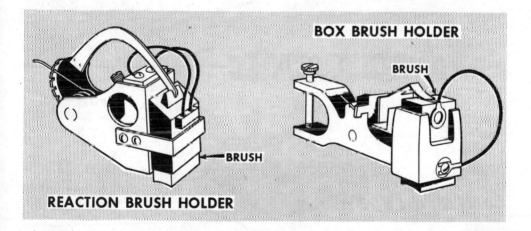

BOX BRUSH HOLDER

BRUSH

BRUSH

REACTION BRUSH HOLDER

The box-type brush holder is made in a great many different styles and shapes. However, each style and shape has the same underlying function: to hold the brush in a sort of box so that the box will be fitted to the commutator in the proper position. A typical box brush holder is shown here.

The box-type brush holder can be set into any position on the motor so that the brush faces in the direction of rotation, against the direction of rotation, or perpendicular to the radius of the commutator. When it is placed at a similar angle, the box type is esssentially the same as the reaction type. However, if the angle of inclination is smaller, care should be taken that the brushes fit snugly within the box, with not too much play. If there is too much play, the brushes will wobble on the commutator, causing poor contact.

Brush Holders — Clamp and Cartridge Types

With the clamp-type brush holder, the brush is clamped in position in the holder by means of a screw and clamp, as shown in the illustration. The brush is held against the commutator by means of a spring. This type of holder is the least used of the three types mentioned, mostly because, at high speeds, the spring reacts too slowly to accommodate the brush to the commutator for proper contact.

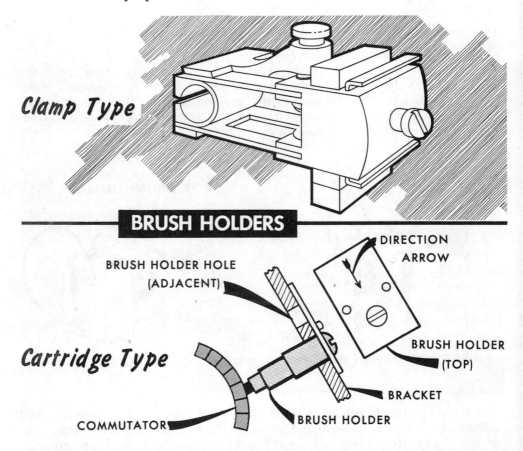

Clamp Type

BRUSH HOLDERS

Cartridge Type

DIRECTION ARROW

BRUSH HOLDER HOLE (ADJACENT)

BRUSH HOLDER (TOP)

BRACKET

BRUSH HOLDER

COMMUTATOR

The cartridge type is usually found in the smaller type of motor and is mounted on a bracket. It is easily removable from the bracket and is usually marked with an arrow to show the direction of rotation. To change the direction, the holder is removed and turned so that the arrow faces in the desired direction, then reinserted. These holders are usually designed to face the commutator at a predetermined angle. For this reason, if reversal is desired, they must be placed in the hole in the bracket adjacent to their original mounting hole when turning is performed. A typical brush holder of the type described is diagrammed here.

Brush Holders — Cartridge and Rocker-Ring Types

Another type of cartridge holder is fixed within its mounting bracket in the motor. This type is semipermanent in that the entire bracket must be removed and reset to a new position to obtain brush shift or a change in direction of rotation. Otherwise, the holders function in the same manner as the cartridge holder just described, except that the brushes ride perpendicular to the radius of the armature, as shown below.

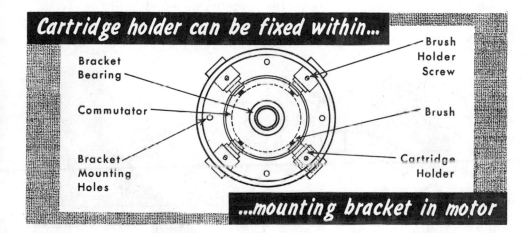

The rocker-ring type of brush holder is usually found in repulsion motors. The holders are fastened to the ring and turn when the ring is turned. The rocker-ring is usually screw-fastened, and shifting is usually accomplished by merely loosening the screws and turning the rocker-ring in the desired direction.

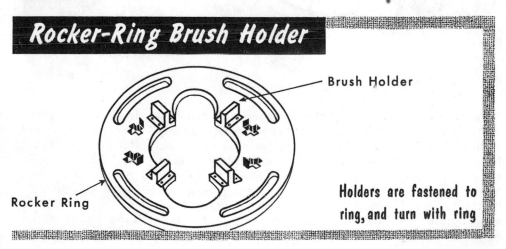

Brush Spacing

The brushes must be properly spaced around the perimeter of the commutator if the motor is to operate satisfactorily. If the spacing is uneven, serious overheating of the armature may occur, as well as possible breakdown of the brushes themselves due to excessive currents. The spacing should be checked every time a brush or brush holder is installed in the motor. A fairly easy way to do this is to wrap a thin strip of white paper around the commutator and mark the position of the brushes on the paper. Care should be taken that the positions are at the same point on each brush. If the spacing between the marks on the paper is uneven, the brush-holder mounting arrangements must be adjusted until the spacing is even.

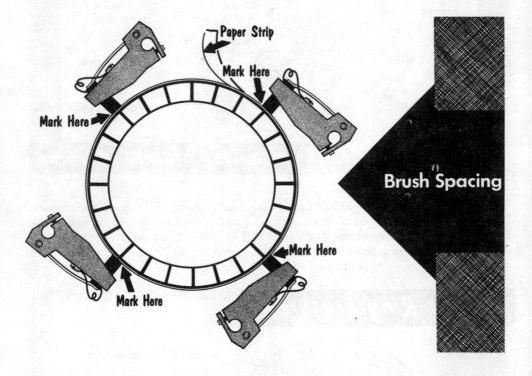

In some motors with only two brushes, the brush spacing should be exactly one pole pitch apart. To check this spacing, some mathematical computations are involved. It requires that the commutator circumference be measured, then be divided by the number of field poles. The result is the distance required between the brushes. If the field poles are cast in the motor frame, this may not be so simple. Sometimes the poles are not correctly spaced. If this condition appears, cutting the tips of the incorrectly spaced pole face shoes will usually correct it. The brushes must be respaced after this operation, as previously described.

Brush Installation

Seating the brushes on the commutator is the process whereby the face of the brush is curved to fit the curve of the commutator. The most common method of performing this operation is by sandpapering. If the brushes are made of a hard graphite, coarse sandpaper may be used for the initial forming. Thereafter, fine sandpaper should be used to obtain a better finish. During the sanding process, which is conducted while the motor is stopped, care should be taken that the brush angle is not changed by the action of the sandpaper and that the brush is not moved back and forth during the sanding. In the final sanding, with fine-grain sandpaper, the sandpaper is drawn across the brush only in the direction the commutator rotates during normal operation.

Following the installation of new brushes on the motor, an important factor to be checked is the pressure applied by the brush to the commutator. This pressure, the spring pressure, should adhere as closely as possible to the manufacturer's recommended pressure. A small spring scale can be used to determine the pressure value. A quick and easy way to use the scale is to place a piece of paper under the brush, lift the brush with the scale, and at the same time gently exert a pulling force on the paper. At the instant the paper begins to move easily out from under the brush, the reading on the scale is taken. If the pressure is not within manufacturing tolerances, the springs must be adjusted until the correct value is obtained. Depending on the type of operation in which the motor is used, recommended pressures vary anywhere from 1¾ to 7 pounds, depending also on the composition of the brushes.

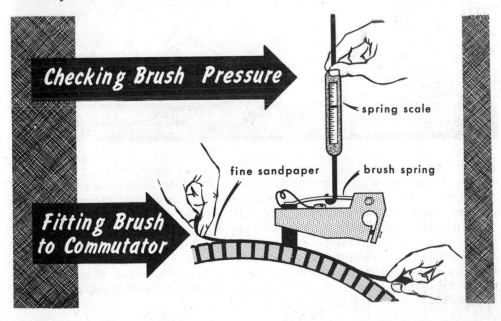

Checking Brush Pressure

spring scale

fine sandpaper

brush spring

Fitting Brush to Commutator

Position of Brushes — Neutrals

Both repulsion- and universal-type motors contain commutators, and the position of the brushes is all important. The determination of the various types of neutrals is discussed as follows: Repulsion-type motors contain two neutral positions: no torque or hard neutral, and soft neutral, which is also a position of no torque. If the brushes are set on the hard-neutral point, the armature does not develop any torque. But if the brushes are shifted in either direction away from this point, torque develops, and the armature rotates in the direction the brushes are shifted. If we turn too far off hard neutral, we encounter the soft neutral, and again no torque. To check whether we are on hard or soft neutral, simply shift the brushes slightly. If the armature turns in the direction of the shift, we are on hard neutral; if the turning is opposite to the shift, we are on soft neutral.

To set the hard neutral, the stator leads must be connected to the line, and the line voltage preferably only 50% of rated input. The brushes must be removed and replaced temporarily with wedge-pointed brushes which will only touch one segment at a time.

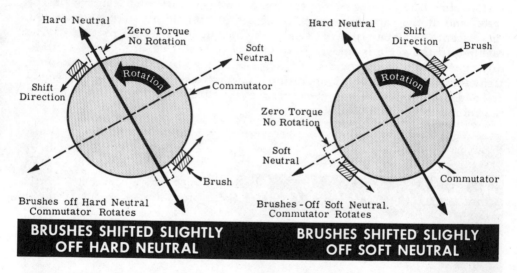

BRUSHES SHIFTED SLIGHTLY OFF HARD NEUTRAL

BRUSHES SHIFTED SLIGHLY OFF SOFT NEUTRAL

Determining hard neutral depends on the type of brush holder used in the motor. If the motor has a rocker-ring type of holder, either the rocker-ring will have three lines scribed on it while one line is scribed on the bracket, or the reverse will be true. Hard neutral is set by setting the center of the three scribed lines exactly in line with the single scribed line. The remaining two lines of the three provide the correct shift for desired direction of rotation. If there are only two lines on the rocker and one on the bracket, hard neutral is set by rotating the rocker until the single line is halfway between the two lines on the rocker. This setting should be made as accurately as possible.

Positions of Neutrals

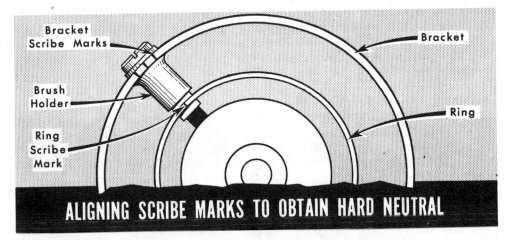

Labels on the diagram:
- Bracket Scribe Marks
- Bracket
- Brush Holder
- Ring
- Ring Scribe Mark

ALIGNING SCRIBE MARKS TO OBTAIN HARD NEUTRAL

On smaller rated motors, using the cartridge type of brush holder with the direction-noting arrow, both the motor frame and the commutator end bracket contain some type of marking to enable setting hard neutral. When the two marks are aligned, we have set hard neutral. If the motor contains no markings to aid us, the brushes must be shifted and the neutral set point determined by a "median" measure. To perform this operation, proceed as follows:

(a) Shift the brush holder mounting bracket until the hard neutral is located. Scribe a mark across both the bracket and motor frame at this point.

(b) Remove, turn, and replace the brushes, to obtain reverse direction of rotation.

(c) Again set the bracket until hard neutral is once more obtained. Scribe a line on the bracket in line with the scribe line put on the motor frame in step (a).

(d) The proper hard neutral is now the median point between the two scribed lines on the bracket. This point is then aligned with the line scribed on the motor frame.

(e) If the present rotation is the desired one, the brush holders may be left in place. If not, they must be removed, turned, and reinserted.

If the brush holders are of the type that is fixed in place in a circular bracket, setting hard neutral is very difficult and is usually not performed unless the armature has been rewound or replaced. The entire bracket must be removed and refastened to a new set of mounting holes, usually found right next to the original mounting holes. If this does not do the trick, the armature must be removed, turned end over end, and reinserted. At the same time, the bracket must be mounted on the other end of the motor. This will reverse the original direction of rotation, which may not be the desired one.

Neutrals of the Concentrated-Pole Universal Motor

In the concentrated-pole universal motor, there are three types of neutrals: mechanical, electrical stalled, and electrical load. All three types have been described in the section under universal motors. The mechanical neutral is the line, drawn perpendicularly through the center of the armature axis, which is exactly 90° away from the line drawn through the centers of

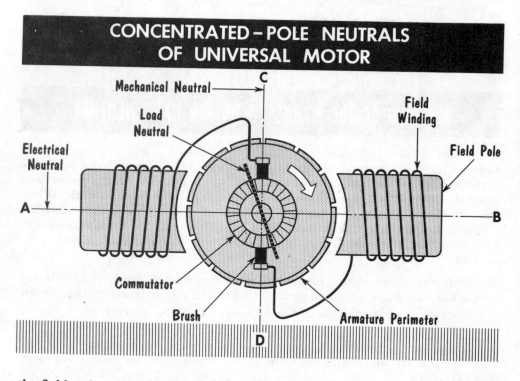

CONCENTRATED-POLE NEUTRALS
OF UNIVERSAL MOTOR

the field poles. This latter line is the electrical stalled neutral. The electrical load neutral lies along a line shifted slightly from the mechanical neutral. To set the electrical load neutral, proceed as follows:

(a) Remove the leads from the brush holders and connect them together.

(b) Jumper the brushes together and connect the motor to a source of approximately half the rated voltage.

(c) Shift the brushes until the armature develops zero torque. This is the electrical stalled neutral point.

(d) Reverse the procedures of steps (a) and (b) and connect the motor both to its load and to its rated power source.

(e) With the motor running, shift the brushes against the direction of rotation until minimum sparking and arcing occur. This is the load neutral point, and the motor should be left in this position.

Neutrals of the Distributed-Field Universal Motor

In the distributed-field universal motor, there are two neutrals: hard and soft, comparable to those in the repulsion motors. To set the brushes on hard neutral, for the nonreversing type, proceed as follows:

(a) Remove the lead from either of the brush holders.

(b) Jumper the two brushes together.

(c) Connect the lead just removed from the brush holder to one side of a source which is no more than half the rated voltage.

(d) Connect the other side of this line input to the main field winding, which is in series with the removed lead.

(e) Shift the brushes until the armature develops zero torque. This is the proper neutral point for the motor.

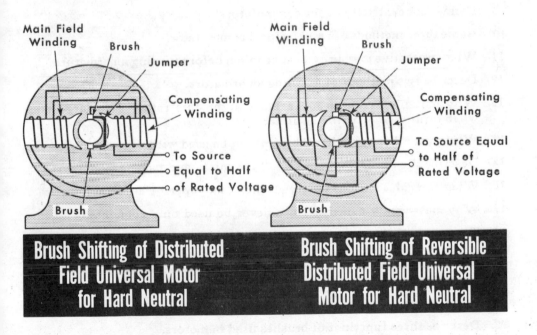

Main Field Winding Brush Jumper Compensating Winding To Source Equal to Half of Rated Voltage Brush

Brush Shifting of Distributed Field Universal Motor for Hard Neutral

Main Field Winding Brush Jumper Compensating Winding To Source Equal to Half of Rated Voltage Brush

Brush Shifting of Reversible Distributed Field Universal Motor for Hard Neutral

If the distributed-field motor has been constructed for reversal operation, the hard neutral may be located as follows:

(a) Disconnect the motor leads from the line input.

(b) Jumper the brushes together.

(c) Take the ends of the filed windings and connect them to a source of low voltage. This source should be no more than half of rated voltage.

(d) Shift the brushes until the armature develops zero torque. This is the proper neutral point for the motor.

CARE AND MAINTENANCE OF MOTORS

QUESTIONS AND PROBLEMS

1. What is the usually recommended brush contact pressure?

2. What instrument measures a running motor's surface temperature?

3. Why should a hand grease gun be used to lubricate ball or roller bearings rather than foot-operated or air-compressor grease guns?

4. Name five general characteristics of a high-grade grease.

5. Where and why should the motor be marked before it is disassembled?

6. What substances can be used for soaking old grease off a bearing?

7. What is the preferred type of oil for relubricating sleeve bearings?

8. What caution must be taken when a motor utilizing the waste-packed type of bearing is placed in operation for the first time?

9. Define "concentricity of the commutator."

10. Name three methods of resurfacing a commutator.

11. What protective measures must be taken before turning an armature?

12. Describe two methods of grinding an armature.

13. Why is it preferable to make small, light cuts rather than large ones when turning or grinding an armature?

14. Why must oil or any other lubricant not be used with a handstone?

15. Describe the condition known as "high mica" and name 3 symptoms.

16. What two rules must be observed when sandpapering a commutator?

17. Why must emery paper or cloth never be used on a commutator?

18. How are invisible scratches removed from a resurfaced commutator?

19. Define the process of undercutting. What is its purpose?

20. Describe the use of a slotting file.

21. What is meant by the term "mica fins"?

22. Describe three functions of brushes used in motors.

23. Name five general categories of brush holders.

24. What are some of the results of uneven brush spacing?

25. Define proper seating of brushes on the commutator.

26. In what direction will the armature rotate if the brushes are shifted in either direction from the hard-neutral point? If the brushes are shifted from the soft-neutral point?

27. At what points of brush setting is there zero torque and no rotation?

Control Devices

In the following pages, we shall deal with the basic means employed to control motor operations. Basically, three types of controls exist in modern motors: manual, automatic, and a combination of both. In the first category are switches of all shapes and sizes; variable transformers, known as Variacs; variable resistors or potentiometers and rheostats; and tapped transformers or chokes. The second category comprises magnetic contactors and their offspring, relays, and thermal devices. The final classification consists of combinations of these two categories, for example, a thermal device used in conjunction with a magnetic device or relay.

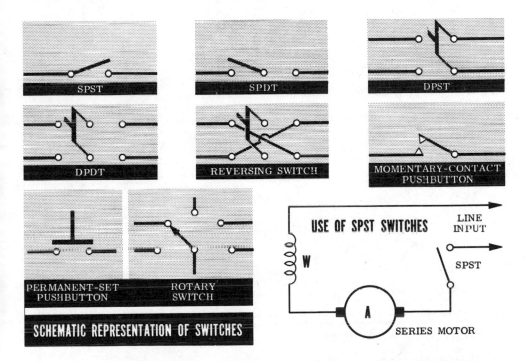

Switches come in many shapes, sizes, and types of operation. However, most switches used in our motor-control circuits are of four or five basic types: single-pole single-throw (SPST); single-pole double-throw (SPDT); double-pole single-throw (DPST); double-pole double-throw (DPDT); reversing, momentary-contact pushbutton; permanent-set pushbutton; and rotary switches. All these switches are illustrated here schematically.

SPST switches are used generally for on-off applications, such as applying or removing source power to the motor. They usually come in toggle form but have been found as knife switches. The momentary-contact pushbutton switch is actually another form of SPST switch.

Switches — SPDT, DPST

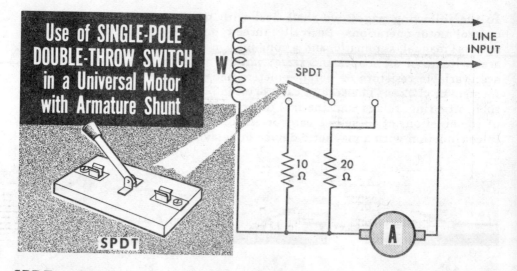

Use of SINGLE-POLE DOUBLE-THROW SWITCH in a Universal Motor with Armature Shunt

SPDT switches are generally used to switch one component into and another out of a circuit at the same time. They come in both the toggle and knife-blade forms, with the toggle type used more often where low voltages appear. The shunt, shown in the figure, is a speed-control device.

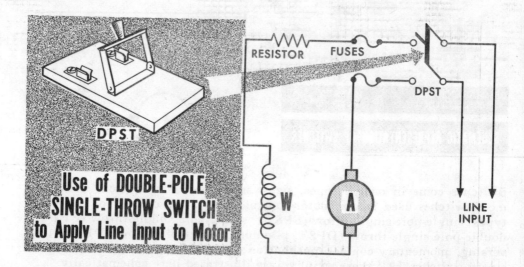

Use of DOUBLE-POLE SINGLE-THROW SWITCH to Apply Line Input to Motor

DPST switches are generally used to connect the line input to the motor, as in the SPST. The DPST is more often used for the line control in d-c applications, especially where starting applications are necessary. The DPST also comes in toggle or knife-blade form.

144

CONTROL DEVICES

The DPDT and Reversing Switches

The DPDT switch, as diagrammed here, is one of the most common of all switches used in motor-control systems. It can usually be found in dual-voltage connections and occurs in both toggle and knife-blade form.

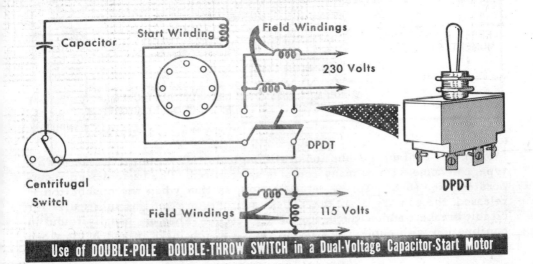

Use of DOUBLE-POLE DOUBLE-THROW SWITCH in a Dual-Voltage Capacitor-Start Motor

As illustrated, the reversing switch simply consists of a DPDT with opposite contacts cross-connected. It is used to reverse connections to or from a circuit. The reversing switch is most often employed in reversible motors and is more commonly found in knife-blade, rather than in toggle, form.

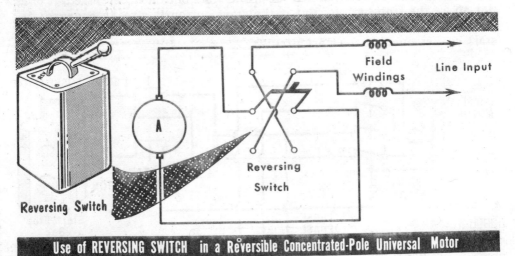

Use of REVERSING SWITCH in a Reversible Concentrated-Pole Universal Motor

Pushbuttons

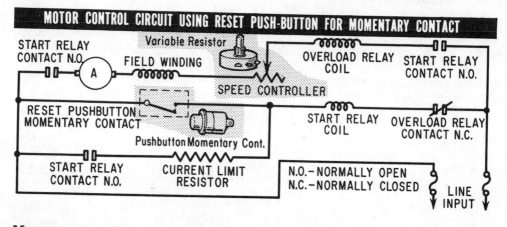

MOTOR CONTROL CIRCUIT USING RESET PUSH-BUTTON FOR MOMENTARY CONTACT

Momentary-contact pushbuttons, although initially shown as an SPST type, can come with as many contacts as required. The basic design incorporates some sort of spring return action, so that when the pushbutton is released, the contact is broken. These switches are used as substitutes for circuit breakers and, as such, are called "resets." They are usually found in conjunction with another automatic device which locks in and holds when the pushbutton is released.

Permanent-set pushbuttons, although shown as merely make-break contact devices, usually come with a double button. That is, as one button is depressed, it makes the circuit, while, when the other is depressed, it breaks the circuit. This is the most common form used in most motor-control circuits. The pushbuttons used are almost always "start" or "stop." The uniqueness of this switch is that when the start button is depressed, it does not affect the stop button. But if the start button is already engaged, depressing the stop button activates a release mechanism, disengaging the start button, opening the start circuit.

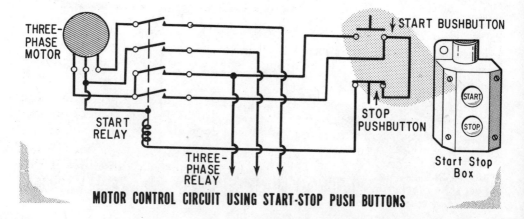

MOTOR CONTROL CIRCUIT USING START-STOP PUSH BUTTONS

146

Other Control Devices

Of the many applications of the rotary switch, the most common is its use with variable-speed and/or reversible motors. The multi-contact type is useful in variable-speed motors, the contacts connecting different voltage sources to the motor. At the same time, these contacts can be used to switch in different motor windings and accomplish reversal action, as shown in the connection diagram.

Variable transformers (Variacs) come in various sizes and ratings and are used chiefly for variation and control of source voltage or voltages. The

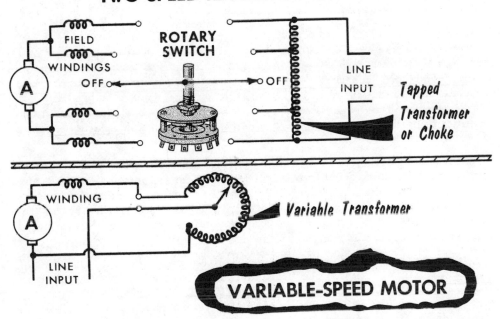

advantage of the Variac is that it can be set at any desired position to account for unexpected surges or drops in the line voltage and to maintain a reasonably constant input.

Variable resistors, or potentiometers and rheostats as they are called, also come in various sizes and ratings and can be used for the control of line-voltage sources, current limiting, and speed control.

Tapped transformers and chokes are essentially the same as the variable transformers in function and application. The one basic difference is that where the variable transformer provides continuously variable voltages, the tapped transformer or choke provides varied fixed voltages.

Automatic Control Devices

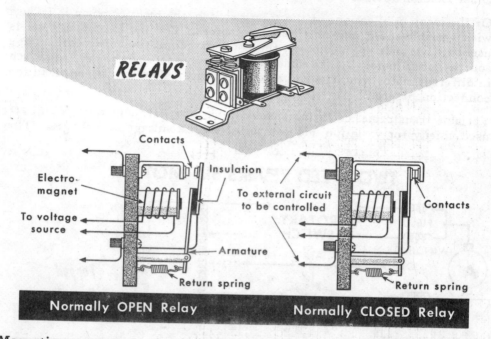

RELAYS

Contacts
Insulation
Electro-magnet
To external circuit to be controlled
Contacts
To voltage source
Armature
Return spring
Return spring

Normally OPEN Relay **Normally CLOSED Relay**

Magnetic contactors and relays are used in all sorts of modern control circuits. Their chief benefit is that they operate automatically with any fluctuation in the circuit which encompasses them. In their major role as protective devices, they are treated in another section of this book, but, since they find widespread use as control devices, they are also included here.

Magnetic contactors and relays are basically similar in their theories of operation but differ in some respects. A contactor may be used to open or close a circuit under normal operating conditions, while a relay may be used to affect one circuit if the relay coil is affected by another circuit, either in normal or failure operation. A magnetized coil operates contacts to open and close them in both contactors and relays. Therefore, for simplicity, we use the term *relay* here to include magnetic contactors.

Relays come with as few as two and as many as 16 or 20 contacts. Each contact is opened or closed by means of magnetism. The relay contains a coil, which, if current passes through it, becomes an electromagnet operating a moving arm. This moving arm can close (make) contacts, or open (break) them, or both (make-break). There are many types of relays: make relays, break relays, make-before-break, break-before-make, time delay, etc. Each of these is some form of automatically operated switch. For example, the start relay shown here is actually a four-pole single-throw switch with the coil thrown in to make the switch action both automatic and simultaneous.

Thermal Devices

Thermal devices, as used in motor-control systems, are usually confined to some type of thermostat. A thermostat is a device which makes or breaks a circuit and whose action is dependent on the surrounding temperature. Quite a few thermostatic units are used as protective devices, but they have found their way into the control field. For this reason, they are discussed under *Control Devices*, as well as under *Protective Devices*.

Thermostats are designed for two basic operations: normally closed-permanently open, or normally open-temporarily closed. In the normally closed-permanently open, the action is fuselike. When the temperature of the surrounding atmosphere exceeds a preset point, the thermostat opens up and must be replaced with a new one when conditions return to normal. This is essentially a protective action.

The normally open-temporarily closed type of thermostat closes when the surrounding temperature exceeds a preset limit. However, once the temperature falls below this point, the thermostat reverts to its normally open position, and the circuit is once more incomplete. Application of this control is found in air conditioners, where the temperature of the area being cooled is the deciding factor as to when the air conditioner is turned on and off. The same principle applies to the operation of oil-burner motors, except that the motor is switched on and off for the opposite reason from the air conditioner.

There are two basic types of temporarily closed thermostats in modern usage; the direct-contact type, and the capillary-bellows type as shown below schematically.

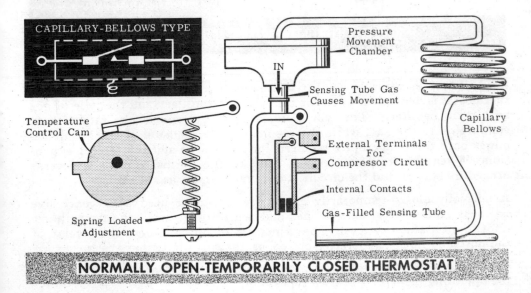

NORMALLY OPEN-TEMPORARILY CLOSED THERMOSTAT

Thermal Devices (contd.)

The direct-contact type is the simpler one. It acts directly on the principle of heat expansion. The contacts are made of a specific metal of known expansion coefficient. The distance between contacts and the expansion coefficient determines the thermostat's operating point. When the temperature drops below the preset level, the contacts contract and reopen the circuit, as shown.

In the capillary-bellows type, the heat-expansion principle is still used but on a gas instead of a solid. Construction of this type consists of a small, metal, gas-filled tube with the capillary attached to an expanding chamber,

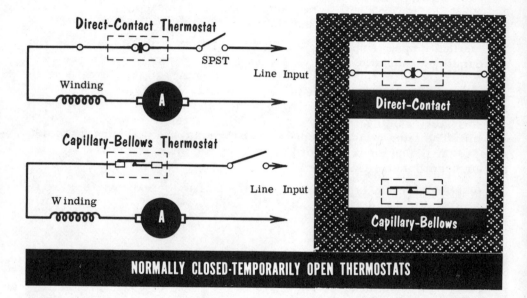

the bellows. A movable contact is affixed to the bellows. Action of the thermostat is dependent upon the metal of the capillary and the type of gas within the capillary. The type of metal used determines the rate of heat conduction to the gas; while the coefficient of expansion of the gas determines both the temperature required for expansion and the rate of expansion. When the temperature falls back to the normal limit, the reverse procedure occurs, and the circuit is once more open, as shown.

A normally closed-temporarily open type would be used where excessive coolness causes the circuit to open and remain open until the temperature rises to a satisfactory level. The principles of operation are the reverse of the normally open-temporarily closed, as shown.

Combinations of both the manual and the automatic control devices utilize the operating principles of both types.

CONTROL DEVICES

QUESTIONS AND PROBLEMS

1. Name the three basic types of controls used presently for motor control. What are some of the controls in each category?

2. Name five basic types of switches used for motor control and an application of each.

3. Explain the difference between momentary and permanent-set pushbutton switches.

4. How does a permanent-set pushbutton usually work?

5. Explain the basic difference between a potentiometer and a Variac.

6. What advantage has the rheostat over the potentiometer?

7. What is the basic theory behind a relay?

8. Name some basic types of relays.

9. What is a thermostat and how does it work?

10. Name the two basic operations of a thermostat. What are the two most common types found in motor-control devices today?

11. Explain fully the theory of operation of the direct-contact type of thermostat.

12. Explain the theory of operation of the capillary-bellows type of thermostat.

Motor Protective Devices

The old adage, "An ounce of prevention is worth a pound of cure," finds application in the field of motor protection devices. This field has grown from a simple fuse to a dazzling selection of devices which make the fuse seem as ancient as the Model T Ford. To attempt to describe and illustrate the various devices used for motor protection would require many more pages than those available to us here. We shall therefore limit ourselves to the basic protective devices, much as we did in the previous section dealing with control devices.

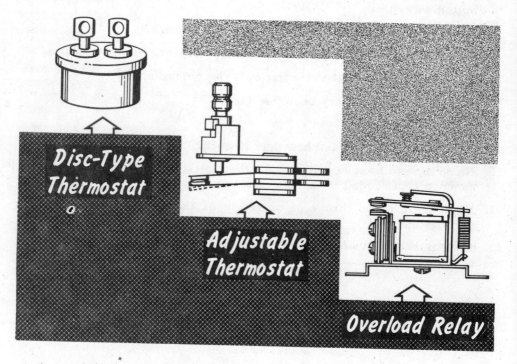

What is a protector? It is a device which is actuated automatically to shut down the motor circuit before too much damage can be done to the motor.

What does a protector look like? Protection devices come in all shapes and sizes, so that no definite picture encompasses them all.

How does a protector operate? Modern protectors work on either of two principles: electromagnetism and heat. Sometimes the protector consists of a combination of both, where a heat-type protector actuates a magnetic-type protector.

The protective devices we shall discuss are included in an overload motor circuit and consist of simple fuses, thermal relays, thermostats, thermally operated disc-type protectors, overload relays, and circuit breakers.

Fuses — Thermal Relays

A simple fuse usually consists of a piece of metal which has a low melting point. Fuses used in our protection circuits usually come in two shapes: the cartridge type and the threaded-screw type. Generally, fuses are used more in the protection of the circuit which provides the motor source voltages rather than in the protection of the motor itself. Naturally, the motor stops if the source of power is cut off.

Both fuses operate on the same principle: a current flowing through a conductor generates heat in that conductor. The greater the current, the more heat is generated. A sudden current surge flowing through the conductor generates tremendous heat sufficient to melt the low-melting-point metal, and the fuse circuit is opened.

Such a fuse is a one-shot affair, and a newer type of fuse has been designed, called the "slow-blow" fuse. The metal of this type fuse is of such composition that a sudden current surge does not cause it to melt unless the surge is of a slightly longer time duration. This type of fuse is used in motors where the starting current is much higher than the running current. It is able to handle the starting current surges.

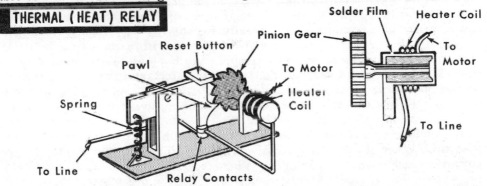

THERMAL (HEAT) RELAY

Solder Film · Heater Coil · Pinion Gear · Reset Button · Pawl · To Motor · Heater Coil · To Motor · Spring · To Line · Relay Contacts · To Line

Overheating of motor causes solder film to melt. This in turn, rotates pinion gear, releasing start button.

The thermal relay is a fairly common protective device, not too often used because of its cost and size. It makes use of a low-melting-point metal, similar to solder. During normal operation, the metal is strong enough to hold a ratchet wheel in place. When an overload occurs, the metal melts, and the ratchet is no longer held fixed in place. With the ratchet free, a reset button held under spring tension is released, and the circuit opens. With the circuit open, the metal cools rapidly, contracts, and once more fixes the ratchet in place. Depressing the reset button now closes the circuit again, while the button is held fast by the immovable ratchet.

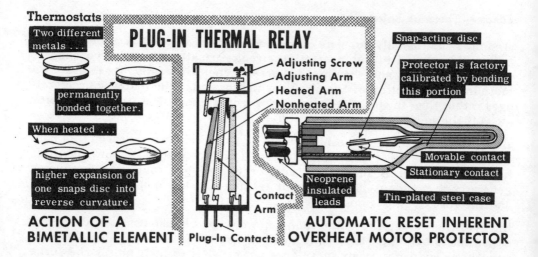

Thermostats

Two different metals . . . permanently bonded together.

When heated . . . higher expansion of one snaps disc into reverse curvature.

ACTION OF A BIMETALLIC ELEMENT

PLUG-IN THERMAL RELAY

Adjusting Screw
Adjusting Arm
Heated Arm
Nonheated Arm

Contact Arm

Plug-In Contacts

Neoprene insulated leads

Snap-acting disc

Protector is factory calibrated by bending this portion

Movable contact
Stationary contact

Tin-plated steel case

AUTOMATIC RESET INHERENT OVERHEAT MOTOR PROTECTOR

Let us explain the function and operation of the thermostat as a motor protector. Two types of thermostats exist in the present-day motor field. The most popular is the snap-acting disc-type thermostat, as shown above, which can be operated again and again, usually for the life of the motor using it. The unit is hermetically sealed and is employed within the motor housing. It responds to the motor winding temperature but only as running protection. Sudden surges of current, such as during start, and the attendant heat, do not affect this unit. Therefore, it must be used in conjunction with devices protecting the starting circuits or those preventing overloads. As shown, the snap-acting thermostatic disc is located in the bottom of the metal enclosure so it can follow the variations in the motor winding temperature. When the disc reaches opening temperature, it snaps into reverse curvature and opens its contacts. When it has cooled to its closing temperature, the disc snaps back automatically, once more closing the circuit. In the wiring diagram shown, the coil is that of a start relay with two normally closed contacts.

The second type of thermostat is similar to the fuses already described and works on the same principle, heat melting metal, but works directly on the heat not on a current flowing through it. It is a one-shot affair in that once the low-melting-point metal melts, the thermostat is rendered useless. This device is not too often used. It finds more applications in controlling the surrounding temperature of the motor rather than the motor circuits. That is why it is often found in fire-alarm circuits. A thermostat can act as a protective device in a typical application where the motor is enclosed in an area where temperature is a problem. Then the thermostat acts through another protective device, usually a relay to cut off the motor when the temperature rise is too high. In the circuit shown, when the thermostat opens, the relay de-energizes, and the motor stops.

Thermally Operated Disc-Type Protectors

One of the most popular of all thermal protective devices is the disc-type illustrated in the cutaway diagram below. Although sometimes called a thermal overload relay, we shall not use the term "relay," since, in the sense we know it, a relay denotes some sort of a winding and associated contacts. This device uses no coil with current flowing through it, but operates on a temperature-affected metal disc. With normal loads, the current through the device and the motor temperature are not enough to trip it. During an overload, the increased current passing through the disc and the more than normal temperature of the motor raise the disc temperature to its opening setting. The disc then snaps to reverse curvature, opening the circuit. This entire action is illustrated on page 154.

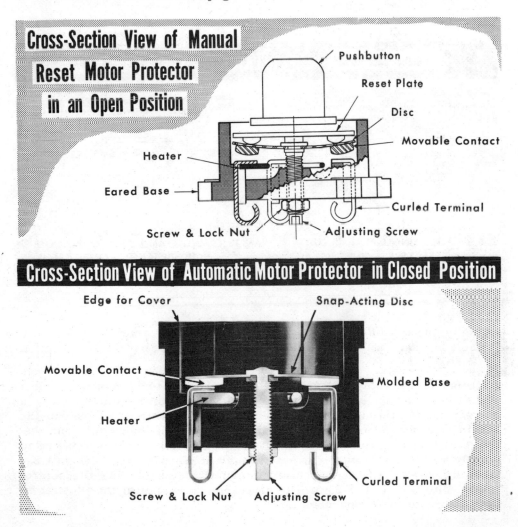

Cross-Section View of Manual Reset Motor Protector in an Open Position

Pushbutton
Reset Plate
Disc
Movable Contact
Heater
Eared Base
Curled Terminal
Screw & Lock Nut
Adjusting Screw

Cross-Section View of Automatic Motor Protector in Closed Position

Edge for Cover
Snap-Acting Disc
Movable Contact
Molded Base
Heater
Screw & Lock Nut
Adjusting Screw
Curled Terminal

Overload Relays and Circuit Breakers

In contrast to the thermal operation principles discussed up to now, we turn to those protective devices which use electromagnetism as their actuating means. Included under this heading are two of the more popular devices in the motor protection field: overload relays and circuit breakers.

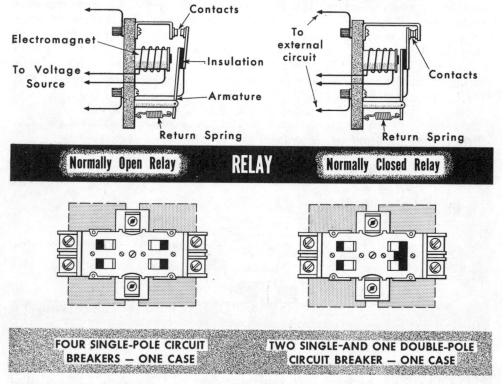

Actually the circuit breaker is in itself an overload relay, but the term "circuit breaker" has become so popular that it has become a classification by itself.

In theory of operation, the overload relay is almost a twin to the relays discussed in the section on Control Devices. That is, it operates by means of a current passing through a coil, causing sufficient electromagnetism in the coil to actuate a moving arm, called an armature. The major difference between the overload and control relays is that the coil of the overload relay usually requires a heavier current through it to make the coil pull the armature down. Although, constructionwise, they look like the control relays illustrated in the Control Devices section, we repeat the illustration here for the reader's convenience.

Overload Relay

The overload relay is designed so that normal motor current does not generate sufficient electromagnetism in the relay coil to operate the relay armature. However, if the motor draws excessive current on either the start or run operation, the field generated in the relay coil is sufficient to actuate the relay armature and pull it down, opening or closing a circuit, depending on whether a normally closed or normally open relay is used.

Usually the overload relay is a locking type, that is, it must be mechanically reset before the circuit can be completed again. The typical operation of both normally closed (N.C.) and normally open (N.O.) relays as motor protectors is diagrammed below.

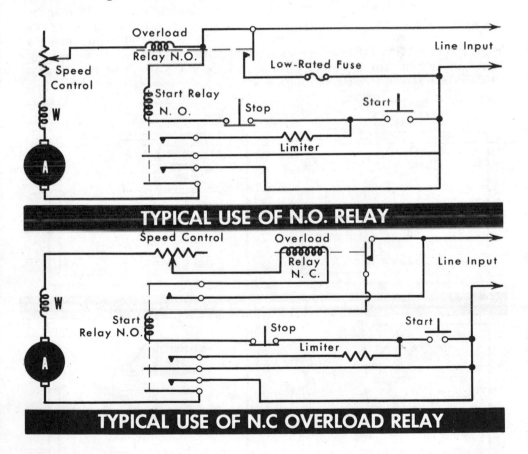

With the N.C. relay, an overload opens the relay, causing it to de-energize and to open the motor circuit. With the N.O. relay, an overload closes the relay, and the line-input current goes through and blows the fuse. The start relay is thus de-energized, and the motor circuit is opened.

The Circuit Breaker

Our final type of motor protector is the circuit breaker. There are three basic types of circuit breakers: the thermal type, thermal-magnetic type, and fully magnetic type.

A thermal circuit breaker responds only to temperature changes. Its principle of operation is based on temperature changes in a bimetal element. Current flows through the bimetal element itself and generates heat. The greater the current, the higher the temperature of the element. The mechanism is set so that the bimetal element bends just enough to open the contacts at a predetermined specified current. A thermal-type circuit breaker is illustrated here.

A thermal-magnetic circuit breaker operates in essentially the same manner as a thermal breaker, except that a magnetic plate is added to speed up the opening of the contacts, without waiting for the bimetal element during a heavy short circuit. Below heavy short circuits, the magnetic force produced by the current is not a factor, and only the heating of the bimetal element operates the contacts. A thermal-magnetic circuit breaker is illustrated here.

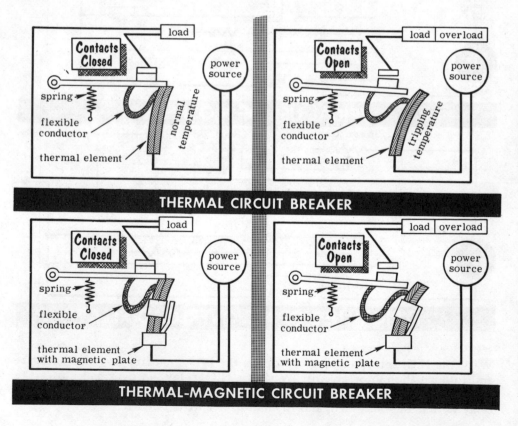

THERMAL CIRCUIT BREAKER

THERMAL-MAGNETIC CIRCUIT BREAKER

Fully Magnetic Circuit Breaker

A fully magnetic circuit breaker responds to changes in current only if they are sufficient to attract an armature by magnetic force. The current rating of a fully magnetic breaker is determined by the number of turns and the wire size of the magnet coil. In the fully magnetic breaker, there is only one actuating element, the magnet coil assembly. Its action is independent of temperature.

Within the magnet coil, an hermetically sealed, nonmagnetic tube contains a movable iron core immersed in silicone fluid. A compression spring normally holds the iron core at the end of the tube, away from the pole face.

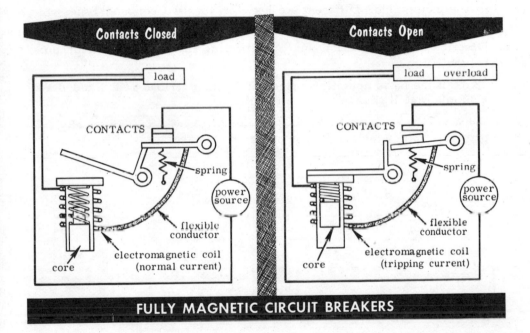

FULLY MAGNETIC CIRCUIT BREAKERS

If the current through the coil is sufficient to draw the movable core into the coil and toward the pole face, the reduction in length of the air gap will produce a greater magnetic pull on the armature and cause attraction.

At normal loads, neither core nor armature moves and nothing happens. At slight overloads, the force of the spring is overcome, and the core moves slowly toward the armature, which is then attracted and opens the contacts. The greater the overload, the faster the core moves through the retarding fluid to open the contacts.

At short-circuit overloads, the magnetomotive force is sufficient to overcome the reluctance of the magnetic circuit at any core position, assuring instantaneous short-circuit tripping.

QUESTIONS AND PROBLEMS

1. What is a motor protector, with reference to the motor field? When does it usually function?

2. Describe the construction and operation of the simple fuse.

3. What is the difference between simple and slow-blow fuses?

4. Describe the operation of a thermal overload relay.

5. Explain the function and operation of the thermostat, with reference to the motor field. Differentiate between the two common types employed.

6. Describe the construction and operation of the disc-type motor protector.

7. Explain the basic operating principle of the overload relay, and draw a simple sketch showing it in a circuit.

8. What are the three basic types of circuit breakers? Explain briefly the principle upon which each type operates.

GLOSSARY

Ampere — A unit of intensity of electrical current produced in a conductor by an applied voltage.

Armature — A special type of rotor.

Bearing — A device used to support the motor shaft, allowing it to rotate smoothly with a minimum of wear and friction.

Bimetallic disc — A disc made up of two strips of dissimilar metals combined to form a single strip.

Breakdown torque — The maximum torque which a motor will develop under increasing load conditions at rated voltage and frequency without an abrupt drop in speed.

Commutator — A device which reverses the connections to the revolving loops (conductors) on an armature.

Cps — Cycles per second.

Efficiency — The ratio of output power (watts) to input power (watts).

Electric motor — A machine which converts electrical energy to mechanical energy.

Electromagnet — A magnet comprised of a coil of wire wound around a soft-iron core. When current is passed through the wire, a magnetic field is produced.

Fractional horsepower motor — NEMA definition of a fractional horsepower motor is a motor built in a frame having a continuous rating of less than one horsepower, open construction, 1700–1800 rpm.

Frequency — The number of complete cycles of current per second taken by alternating current.

Ground (motor) — Any point on a motor component where the ohmic resistance between the component and the motor frame is one megohm or less.

Horsepower — A measure of the time-rate of doing work, defined as the equivalent of raising 33,000 pounds one foot in one minute. One hp equals 746 watts.

Induced current — The electric current produced by moving a conductor in a magnetic field.

Insulation — A material having a relatively high resistance.

Megohm — One million ohms.

NEMA — National Electrical Manufacturers' Association. This organization establishes certain voluntary standards relating to motors: such as operating characteristics, terminology, basic dimensions, ratings, and testing.

No-load speed — The speed reached by the rotor or armature when it rotates freely within its bearings with no restraining force (load) attached.

Ohm — A unit of electrical resistance of a conductor.

Overload protector — A device affected by an abnormal operating condition which causes the interruption of current flow to the device governed.

GLOSSARY

Percentage slip — The ratio of synchronous speed minus actual speed divided by the synchronous speed of an induction motor. This is expressed mathematically as:

$$\text{Percentage slip} = \frac{(\text{Synchronous speed} - \text{Actual speed}) \times 100}{\text{Synchronous speed}}$$

Power factor — The figure which indicates what portion of the current delivered to the motor is used to do work.

Pull-in torque — The maximum torque at which an induction motor will pull into step (2 — 5% below synchronous speed).

Pull-up torque — The minimum torque developed by an induction motor during the period of acceleration from rest to full speed.

Relay — A device that operates by a variation of a condition to effect the operation of other devices in an electric circuit.

Rotor — The rotating section which rotates within the stator of a motor. The rotor consists of a core, windings, and a shaft.

Rpm — Revolutions per minute.

Running torque — The torque or turning effort determined by the horsepower and speed of a motor at any given point of operation.

SAE — Society of Automotive Engineers.

Short — Any two points of a motor where there is zero, or extremely low, resistance between them or between two motor components.

Starting torque — (Sometimes called locked rotor torque.) The amount of torque produced by a motor as it breaks the motor shaft from standstill and accelerates.

Stator — The stationary section within which the rotor (rotating section) of a motor rotates. The stator consists of three parts: an outer frame, a core, and windings. The stator windings and core form the electromagnet which produces the magnetic field within which the rotor turns.

Synchronous speed — The constant speed to which an a-c motor adjusts itself, depending on the frequency of the power supply and the number of poles in the motor.

Thermostat — An instrument which responds to changes in temperature to effect control over an operating condition.

Voltage — A unit of electromotive force. It is a force which, if applied to a conductor, will produce a current in the conductor.

PICTURE CREDITS

INDEX

INDEX

INDEX

fusible, 154
snap-acting disc, 154, 155
Three-phase motors, 10, 74, 75
    questions and problems, 85
    reversal, 81
    squirrel-cage, 78, 79
    starting small synchronous motor, 77
    synchronous, 76
    troubleshooting, 82–84
    wound-rotor, 80
Troubleshooting
    capacitor motors, 30, 31
    repulsion motors, 43–45
    shaded-pole motors, 52–54
    split-phase motors, 19–21
    three-phase motors, 82–84
    universal motors, 70–72
Two-phase motors, 7
Two-speed capacitor motor, 29
    motor connections, 29
    reversal, 29
Two-value capacitor motor, 27

Universal motors, 55, 56
    commutation, 57
    commutator action, 58
    concentrated-pole, 61–63
    distributed-field compensated, 68, 69

no-load speed control, 64–66
    questions and problems, 23
    speeds, 59
    troubleshooting, 70–72
    types of, 60

Windings, 94
    chain, 101
    concentrated, 102
    definitions of terms, 94
    distributed, 101
    double re-entrant, 98
    Gramme-ring, 95
    half-coil, 103
    hand, 103
    lap, 96
    lap multiplex, 97, 98
    lap simplex, 96
    mold, 103
    single re-entrant, 98
    skein, 102
    types, 95
    wave, 99
    wave multiplex, 100
    wave simplex, 99
    whole-coil, 103
Wound-rotor motor, 80